T0205586

# Springer Theses

Recognizing Outstanding Ph.D. Research

## Aims and Scope

The series "Springer Theses" brings together a selection of the very best Ph.D. theses from around the world and across the physical sciences. Nominated and endorsed by two recognized specialists, each published volume has been selected for its scientific excellence and the high impact of its contents for the pertinent field of research. For greater accessibility to non-specialists, the published versions include an extended introduction, as well as a foreword by the student's supervisor explaining the special relevance of the work for the field. As a whole, the series will provide a valuable resource both for newcomers to the research fields described, and for other scientists seeking detailed background information on special questions. Finally, it provides an accredited documentation of the valuable contributions made by today's younger generation of scientists.

## Theses are accepted into the series by invited nomination only and must fulfill all of the following criteria

- They must be written in good English.
- The topic should fall within the confines of Chemistry, Physics, Earth Sciences, Engineering and related interdisciplinary fields such as Materials, Nanoscience, Chemical Engineering, Complex Systems and Biophysics.
- The work reported in the thesis must represent a significant scientific advance.
- If the thesis includes previously published material, permission to reproduce this must be gained from the respective copyright holder.
- They must have been examined and passed during the 12 months prior to nomination.
- Each thesis should include a foreword by the supervisor outlining the significance of its content.
- The theses should have a clearly defined structure including an introduction accessible to scientists not expert in that particular field.

More information about this series at http://www.springer.com/series/8790

Valerio Rizzi

# Real-Time Quantum Dynamics of Electron-Phonon Systems

Doctoral Thesis accepted by
the Queen's University Belfast, Belfast, UK

 Springer

*Author*
Dr. Valerio Rizzi
Institute of Computational Sciences
Università della Svizzera Italiana
Lugano, Ticino, Switzerland

and

Department of Chemistry and Applied
  Biosciences
ETH Zürich
Zürich, Switzerland

*Supervisors*
Prof. Tchavdar N. Todorov
Atomistic Simulation Centre
Queen's University Belfast
Belfast, UK

Prof. Jorge J. Kohanoff
Atomistic Simulation Centre
Queen's University Belfast
Belfast, UK

ISSN 2190-5053          ISSN 2190-5061   (electronic)
Springer Theses
ISBN 978-3-030-07169-1          ISBN 978-3-319-96280-1   (eBook)
https://doi.org/10.1007/978-3-319-96280-1

*Truth (...) is much too complicated to allow anything but approximations.*

—John von Neumann, *The Mathematician*

# Supervisors' Foreword

The industrial progress of humanity has been driven by learning to control and use energy, sometimes for good aims, sometimes not. Most of our bills—from electricity and heating to travel and communications—are ultimately about energy. A kettle converts energy from electric current into heat to make tea. The same principle enables irradiated metal nanoparticles to generate heat in a targeted way, enabling novel thermal cancer therapies. Current drives electric motors in the household. In atomic wires, it can also drive a nanomotor.

We live in an age of learning to harness energy at the molecular scale. To achieve this we need an understanding of the underlying dynamics, via state-of-the-art experiments and computer simulation. This requires a paradigm shift in simulation that transcends *atomic motion*. We need to simulate the coupled dynamics of the constituent *electrons and nuclei*, at the most difficult middle ground between the macroscopic and atomic limits—the mesoscale.

This regime brings together emerging theoretical, computational and experimental breakthroughs, requiring the simulation of electron-nuclear dynamics in the middle, where experiment and theory are converging across molecular electronics and thermoelectric phenomena, radiation damage in materials and biological systems, hot electrons in photovoltaic devices, photo-electrochemistry and nanoplasmonics.

The fundamental process controlling energy conversion at the molecular scale is the *electron–phonon interaction*. This is how electrons and atomic motion exchange energy and momentum. This maintains thermal equilibrium, enabling stable interatomic bonds to form. It drives thermalization when electrons and vibrations are driven *out* of equilibrium, and the larger the initial excitations the more violent the electron–phonon response. We can see and feel, and even hear and smell, the effects of current flow in a lightbulb. In atomic wires, the current density exceeds that in a lightbulb by 5–7 orders of magnitude.

This vast variation occurs also in the timescales for electron–phonon processes. In a lightbulb, the initial transient lasts a fraction of a second. In a photo-excited molecule, violent electron–phonon processes take place on the femtosecond scale. The variation occurs also in system sizes. Electron–phonon scattering generates a signal in transmission, paving the way for inelastic current–voltage spectroscopy, and this has been done for a diatomic molecule between two electrodes. Even a single dynamical atom in a nanowire can operate as a mini-inelastic resistor.

This huge variation in physical conditions is one reason why there is no single framework for electron–phonon problems. The other is that electron–phonon interactions are a tough *many-body quantum problem*, even without the added difficulty of electron–electron interactions. This is what sets aside the mid-range. At the atomic scale, many-body methods, notably Non-equilibrium Green's Functions, are feasible and have been used with much success. The other end—macroscopic systems—allows key simplifications: departures from equilibrium are weak, while coherent scattering can often be neglected; this makes the linearized semi-classical Boltzmann equation a robust framework. In the middle, neither the highly accurate but expensive, nor the tractable but simplified work well. In addition, extra complexity arises, especially the need for *simultaneous coupled electron–phonon dynamics*, under violent departures from steady-state conditions.

This Ph.D. thesis tackles precisely that problem. It formulates a methodology for real-time coupled quantum dynamics of electrons and phonons in nanostructures, both isolated and open to an environment, including a system of external reservoirs to generate and maintain current flow. It then applies this technique to both fundamental and practical problems, of relevance, in particular, to nanodevice physics, laser–matter interaction and radiation damage in living tissue.

It is a pleasure to wish the reader an enjoyable and fruitful journey through these difficult but important and exciting developments.

Belfast, UK                                                              Tchavdar N. Todorov
July 2018                                                                    Jorge J. Kohanoff

# Abstract

Today the relevance of non-adiabatic phenomena at the atomic scale is ever more significant, as experiments can probe into the microscopic world at ultrafast time-scales. In radiation damage, there is a clear need for new radiation sturdy materials and for innovative ways to selectively damage cancerous cells. An improved understanding of the microscopic dynamics of electrons and atoms can offer critical insights, but the existing simulation methods are either too simplified to describe the full quantum problem or too complex to be able to simulate the large systems required.

The novel quantum method that is presented here can describe explicitly an electronic system in real time together with the phonons. It can model out-of-equilibrium systems that mutually exchange energy. Examples of applications include Joule heating, inelastic electron tunneling and equilibration of hot electrons in irradiated materials. A scalable parallel code has been developed which can simulate multi-electron and multi-phonon systems in a wide range of time-scales: from the electronic attosecond to the atomic picosecond.

Relevant results include the thermalization of out-of-equilibrium electrons and phonons in real time without the need for an empirical thermostat, even in a situation of full electronic population inversion. A water chain system is simulated and phonon-assisted electron injection processes are investigated, highlighting energy ranges where the inelastic exchanges lead to a dramatic cooling or heating.

It is hoped that this method and its applications will serve as a stepping stone for future simulations on more complex systems and provide inspiration for fulfilling ever more daring and wide-reaching goals.

# Acknowledgements

First and foremost, I would like to express my deepest gratitude to my supervisors Tchavdar and Jorge for their constant support and endless patience. While enduring my questions and clarifying my doubts, they have always provided me with an aim and a sense of direction. I really thank Alfredo for being an inextinguishable source of inspiration and for granting me the possibility to collaborate with him in California. I am grateful to the Leverhulme Trust for funding this project under grant RPG-2012-583.

I thank all my good friends that were close to me during these amazing 4 years, especially Federica and Antonio. All the adventures that we went through made my time in Belfast fly. I thank my family that has always supported me from the other side of the sea. At last, I want to thank my wonderful wife Canan whose presence has made each of my days with her brighter and happier.

Valerio Rizzi

# Contents

# Abbreviations

| | |
|---|---|
| 2P | Two-point |
| 2TM | Two-temperature model |
| Å | Ångström |
| a.m.u. | Atomic mass unit |
| as | Attosecond |
| BOA | Born–Oppenheimer approximation |
| CEID | Correlated electron-ion dynamics |
| DEA | Dissociative electron attachment |
| DFT | Density functional theory |
| DM | Density matrix |
| DNA | Deoxyribonucleic acid |
| DOF | Degrees of freedom |
| ECEID | Effective correlated electron-ion dynamics |
| ED | Ehrenfest dynamics |
| EEP | Excess electron population |
| EMFP | Electron mean free path |
| EOM | Equations of motion |
| EPI | Electron–phonon interaction |
| ESP | Electronic stopping power |
| eV | Electronvolt |
| FCB | First conduction band |
| FD | Fermi–Dirac |
| FGR | Fermi golden rule |
| fs | Femtosecond |
| GF | Green's function |
| HML | High mass limit |
| HO | Harmonic oscillator |
| HOWF | Harmonic oscillator wavefunction |
| K | Kelvin |
| keV | Kiloelectronvolt |

| | |
|---|---|
| LEE | Low-energy electron |
| MD | Molecular dynamics |
| MeV | Megaelectronvolt |
| NEGF | Non-equilibrium Green's functions |
| OB | Open boundaries |
| OOE | Out-of-equilibrium |
| PKA | Primary knock-on atom |
| ps | Picosecond |
| SCB | Second conduction band |
| TB | Tight binding |
| TDDFT | Time-dependent density functional theory |
| TDTB | Time-dependent tight binding |
| TH | Top hat |

# Chapter 1
# Introduction

Imagine an everyday situation such as preparing a tray of lasagne, placing it in an oven and turning the switch on. This simple action triggers an electric current to flow into a resistor and heats it up, making it glow red hot and emit infrared radiation. The radiation increases the temperature of the air in the oven and, finally, cooks the lasagne. The heat generated by an electric oven relies upon a fundamental process in solid state physics: electron–phonon interaction.

The same physical phenomenon takes place in the smallest conductor in nature, a molecular nanowire, where the effect of current-induced heating can be extremely intense [1, 2]. This process is called Joule heating and can be visualized by imagining electrons as point particles speeding past the surrounding heavy nuclei [3]. Some electrons collide and bounce off the nuclei, depositing energy into atomic vibrations, the phonons. Electron–phonon scattering is the mechanism that makes an electric current convert electrical energy into heat, which is characterised in terms of atomic vibrations.

The coupled dynamics of electrons and nuclei [4, 5] is central to many problems in physics, chemistry and materials science: nanoscale electronics, photochemistry and photocatalysis, matter in laser fields. Radiation damage is one of the most relevant and complex fields of study because of the wide range of size and time-scales that it involves. During a typical radiation event, an energetic particle clashes against a material and, while slowing down, releases heat causing damage [6–8].

Throughout the process, electrons are heated to high temperatures and can transfer energy back into the atomic motion via electron–phonon interaction. Phonon emission is a key element in the coupled electron–nuclear dynamics. Electrons are the main heat carrier in a metal, so their dynamics strongly determines the cooling down of the material and the subsequent formation of damage. Radiation damage assessment and its prevention are topics of central interest for industries. Understanding the extent of heat removal and determining the area where energy is deposited would provide significant advancements in the field.

At a biological level, when living cells sustain a radiation event, water may be ionised and emit secondary electrons. These electrons diffuse, while interacting with vibrations in nearby molecules and can ultimately reach DNA. Depending on their

© Springer International Publishing AG, part of Springer Nature 2018  
V. Rizzi, *Real-Time Quantum Dynamics of Electron-Phonon Systems*,  
Springer Theses, https://doi.org/10.1007/978-3-319-96280-1_1

energy spectrum and the position of impact, the electrons can damage DNA, causing strand breaks and possibly leading to cellular death [9–11]. The inelastic propagation of electrons in water is a significant phase in biological radiation damage [12] and is not yet completely understood. It is generally modelled as a classical random walk.

An electron–atom collision picture helps to understand the problem in a classical framework, but atoms and electrons are not classical objects. As distances and times shrink, the quantum properties of matter become more and more relevant until, at the nanoscale, they become essential to describe electron–nuclear correlated processes. To probe matter experimentally on such small scales is no easy task. The technical advances in lasers allow to delve into nanomaterials with a remarkable level of detail [13]. Photoelectron spectroscopy experiments and especially the upcoming free electron lasers can supply an unprecedented amount of information about the distribution of out-of-equilibrium electrons and their dynamics on ultrafast timescales. The coherent photon pulses of a free electron laser can have wavelengths of less than 1 Å with time durations of about 1 fs, and will allow to gather state-of-the-art information about the atomic world and its dynamics [14].

Experiments, however, lack the opportunity of controlling and fine tuning all the properties of the probed systems. A tool that can work hand in hand with experiments, offering predictive power while providing total control on the systems, is computer simulation. The possibility of following the journey of an electron, from emission to thermalization, is an exciting prospect for computer simulations, albeit extremely complex. Today, the ever increasing computer power allows the study of larger and more complex systems and makes the development of new ambitious methods meaningful.

Exact simulations of many-body problems soon run into an exponential scaling wall: the storage of the fully correlated information about systems even with very few degrees of freedom would require an enormous amount of computer memory. Most simulations inevitably rely on assumptions that lead to models that can explain or predict only some aspects of a problem. The aim of this thesis is to tackle the quantum-mechanical problem of the propagation of excited electrons in real time through materials and their gradual de-excitation by inelastic electron–phonon scattering.

One of the first studies of the problem of the propagation of out-of-equilibrium electrons in the presence of phonons dates back 50 years [15] and dealt with deviations from Ohm's law in metals. The authors claimed that, in the presence of strong currents, the temperature of the electron subsystem would be significantly higher than that of the lattice and devised heat transfer equations in various limits of temperature differences. Such a programme would be carried out in different forms and with various improvements over the decades [16] and is now known as the two temperature model. This model is still successfully used today [17], but is inadequate to describe the problem at a microscopic level.

Molecular dynamics simulations [18] describe atomic motion microscopically and have to rely upon approximations to include the role of the electrons. A standard approach uses the Born-Oppenheimer approximation which exploits the large mass difference between nuclei and electrons. As the electrons are much lighter than the nuclei, their dynamics occurs on a faster timescale. Therefore, as the nuclei move, it is assumed that the electronic subsystem equilibrates instantaneously to its ground state. For any set of nuclear positions, the electrons produce an effective potential that controls the ensuing nuclear dynamics. Nevertheless, methods based on the Born-Oppenheimer approximation are inadequate for simulating problems where the mutual interaction between nuclei and electrons play a significant role.

One of the earliest attempts to include inelastic energy losses in radiation cascade simulations portrayed the effect of electronic excitations as a drag force proportional to the velocity of the projectile [7]. This friction force was implemented as a random force representing the excited electrons hitting the nuclei. Methods with a higher level of sophistication were developed over the years with the inclusion of an electronic temperature field [19], however such approaches are intrinsically limited by their implicit treatment of electrons.

Methods for describing mutually interacting electrons and nuclei are referred to as non-adiabatic dynamics [20, 21]. The simplest and one of the most widely used is based on the Ehrenfest approximation [22], where the quantum dynamics of electrons is coupled with Newtonian equations of motion for the nuclei. The method employs a mean-field approach to describe the force exerted by the electrons on the nuclei and this causes an imbalance in the description of energy flow. Ehrenfest dynamics manages to capture the flow of energy from excited nuclei to the electrons, but it fails to describe the reverse process where excited electrons lose energy to the nuclei through spontaneous phonon emission [23].

To visualize an analogous asymmetric behavior in an everyday situation, one can imagine an object immersed in a fluid. The fluid would slow down a fast moving object and this effect can be described by considering the fluid as a continuous entity. On the other hand, the warming up of a cold object surrounded by a hot fluid cannot be understood through a mean field approach. It a has to be modelled by considering the molecular collisions of the fluid components on the cold object. The weakness of Ehrenfest methods lies in not being able to capture the individual electronic collisions that lead to nuclear heating.

To go beyond the Ehrenfest approximation, a number of methods have been created, such as Correlated Electron-Ion dynamics (CEID) [24, 25] or the Bonca-Trugman method [26], but none is ideal for the type of problems that we wish to tackle in this work and their scale. The challenge in capturing quantum electron–phonon dynamics is twofold: enough physics must be included for a correct description of the mutual energy exchanges, but substantial approximations must be made for achieving efficiency in simulations of extended systems. Problems such as the dynamics of electronic thermalization during a radiation event or the inelastic diffusion of excited electrons hinge upon both aspects. While Ehrenfest dynamics is too simplified and cannot capture the crucial phenomenon of spontaneous phonon emission, CEID and more advanced methods don't scale well with an increasing system size.

In this thesis, we address this need for a new method and present an efficient and scalable method that treats simultaneously coupled systems of electrons and phonons at the mesoscale. The method describes the non-adiabatic dynamics of electrons and oscillators in a number of out-of-equilibrium situations and can display phonon assisted electron transfer in real time. We developed a parallel code to efficiently simulate electron–phonon systems over a wide range of timescales that spans from the electronic attosecond to the picosecond typical of atoms.

The potential impact of the method involves both the experimental and theoretical communities. Experimentally it can provide insights into the electron dynamics, leading to a better understanding of key inelastic processes in systems under irradiation. From the theoretical point of view, it can pave the way to a class of new efficient codes to model non-adiabatic physical situations.

Here follows an outline of this work.

**Thesis outline**

**Physical motivation**    We provide a context and a motivation for this thesis by describing physical problems where the interaction between out-of-equilibrium electrons and phonons is relevant. We first describe the multiscale field of radiation damage. After showing the phases of a radiation cascade in a metal, we investigate the role of electrons, both as an energy loss mechanism and a factor to take into account for the final material damage. Then we review radiation damage in biological systems, focusing on how secondary electrons can damage DNA according to their energy distribution and highlighting the necessity of inelatic simulations that can track the electronic dynamics in a biological environment. Next, we examine the field of ultrashort excitations of metals where the use of femtosecond lasers and the advancements in pump and probe techniques have allowed to explore highly excited electronic states and to track their dynamics.

**Simulating electrons and phonons: effective temperature methods**    Here we discuss methods that treat electrons implicitly in the study of coupled electron–phonon problems. The assumption underlying these methods is that electrons equilibrate much faster than the atomic system. Thus, at atomic timescales, their explicit fast-varying dynamics can be replaced by a slowly-varying effective electronic temperature. We offer an overview of the works that devised this approach, called the two-temperature model, and applied it to radiation damage problems and laser excited matter. Then we examine the field of molecular dynamics coupled to the two temperature model, starting from the Caro-Victoria model, where electrons are implicitly represented as Langevin thermostats, and ending with inhomogeneous models that discretize space for describing areas of different electronic temperature.

**Simulating electrons and phonons: atomistic methods**    We examine methods that take into account the electrons explicitly. Among the non-adiabatic methods that go beyond the limitations of the Born-Oppenheimer approximation, we consider Ehrenfest dynamics, its application in Time-Dependent Density Functional Theory and its limitations. Then we introduce a method that goes beyond Ehrenfest dynamics by perturbatively expanding the motion of atoms around their mean position, Correlated Electron-Ion dynamics (CEID) in its different versions. At last, we describe the Bonca-Trugman method, an essentially exact method applicable to electron–phonon model systems with a small number of degrees of freedom.

**The ECEID method** In this chapter we describe the method that was developed in this thesis: effective correlated electron-ion dynamics (ECEID). After introducing the model Hamiltonian, we go through the derivation, providing first an exact form of important quantities and then introducing approximations, until a system of equations of motion is reached. After a comparison with Ehrenfest dynamics, we write down the equations of motion in one-electron form and show that they conserve the total energy of the system. We proceed by applying open boundaries to ECEID: an electron injection and extraction mechanism. We outline the implementation of the method in a FORTRAN code.

**ECEID validation** The capabilities of ECEID are explored by comparing it to an exact simulation on small systems made of a few electronic levels and one or two phonons. These systems provide a controlled environment for checking the extent of the ECEID approximations. We study different regions of parameter space and try to provide explanations for the agreement (or the lack of it) between methods. Then we test the open boundaries on one dimensional metal chains by reproducing results from previous work and by recovering the Landauer conduction picture in a well defined limit. A microscopic Ohm's law is tested, where the resistivity of metal chains of a varying length is measured and compared with a perturbative result. Then the role of onsite disorder is explored. At last, we perform code performance tests, focusing on the scaling with the number of electronic sites and oscillators.

**Thermalization with ECEID** We apply ECEID to the problem of thermalization, following our recent paper [27]. We introduce a definition of a temperature-like parameter for out-of-equilibrium electrons and compare its time evolution with the vibrational one. We show the ability of ECEID to thermalize electrons and phonons starting at different temperatures. We simulate a complete electronic population inversion and track its equilibration with a phonon bath, while measuring the heat exchange between subsystems. A kinetic model based on rate equations is employed to rationalize the levels' dynamics.

**Electronic transport in water** Transport properties of water molecules and chains are studied, with a focus on phonon assisted processes. Part of this chapter has recently been published [28]. First, a simple tight-binding model is employed to describe a single water molecule with two modes corresponding to a symmetric and an antisymmetric stretching of the hydrogen bonds. The molecule is connected to metal leads and its elastic transport features from frozen phonon Green's function calculations are compared with ECEID results, that include inelastic effects. In the phonon's high mass and low frequency limit, ECEID converges to the elastic case. Then a water chain is built from DFT calculations that determine the equilibrium geometry and the collective modes. We include a high energy phonon and inject Gaussian electron pulses in the system, investigating their dynamics. We observe phonon assisted heating or cooling depending on the pulse energy. The charge deposited into the water generates a polaron with a lifetime that depends on the water energy levels involved.

**A new development: ECEID xp**   We derive a more general version of the method that includes a time-dependent parameter that can be associated with the motion of the oscillators centroids. We call this new development ECEID xp and observe that, in the limit of frozen centroids, it reduces to ECEID, with the advantage of a more concise set of equation of motion and a more straightforward derivation. A condition is proposed to infer an equation of motion for the centroids and a test case is shown.

**Appendix: Electronic operators in ECEID: from Many-body to Single body** This appendix contains a schematic derivation of the projection procedure of the general many-body ECEID equations of motion into the single-body ones, that are used in the simulations.

**Appendix: Open Boundaries in ECEID**   The formulation of the Open Boundaries method is shown. The method allows electron injection and extraction in a system by coupling to external reservoirs. The additional terms needed in ECEID to allow injection and extraction are presented.

**Appendix: An alternative water chain**   A water chain model with simpler intermolecular hoppings than the ones in Chap. 8 is used to inject electrons both as a time dependent pulse and a constant stream. Long lived excitations form because of the chain band structure.

**Appendix: Beyond the double (de)excitation approximation**   A methodological improvement is proposed where one of the ECEID approximations is not invoked.

# References

1. McEniry, E., T. Frederiksen, T. Todorov, D. Dundas, and A. Horsfield. 2008. Inelastic quantum transport in nanostructures: The self-consistent Born approximation and correlated electron-ion dynamics. *Physical Review B* 78 (3): 035446. https://doi.org/10.1103/PhysRevB.78.035446.
2. Fangohr, H., D.S. Chernyshenko, M. Franchin, T. Fischbacher, and G. Meier. 2011. Joule heating in nanowires. *Physical Review B* 84 (5): 054437. https://doi.org/10.1103/PhysRevB.84.054437.
3. Ashcroft, N.W., and D.N. Mermin. 1976. *Solid State Physics*. Philadelphia: Saunders College.
4. Galperin, M., M.A. Ratner, and A. Nitzan. 2007. Molecular transport junctions: Vibrational effects. *Journal of Physics: Condensed Matter* 19: 103201. https://doi.org/10.1088/0953-8984/19/10/103201.
5. Giustino, F. 2017. Electron–phonon interactions from first principles. *Reviews of Modern Physics* 89 (1): 015003. https://doi.org/10.1103/RevModPhys.89.015003.
6. Gibson, J., A. Goland, M. Milgram, and G. Vineyard. 1960. Dynamics of radiation damage. *Physical Review* 120 (4): 1229–1253. https://doi.org/10.1103/PhysRev.120.1229.
7. Caro, A., and M. Victoria. 1989. Ion-electron interaction in molecular-dynamics cascades. *Physical Review A* 40 (5): 2287–2291. https://doi.org/10.1103/PhysRevA.40.2287.
8. Correa, A.A., J. Kohanoff, E. Artacho, D. Sánchez-Portal, and A. Caro. 2012. Nonadiabatic forces in ion-solid interactions: The initial stages of radiation damage. *Physical Review Letters* 108 (21): 213201. https://doi.org/10.1103/PhysRevLett.108.213201.
9. Boudaïffa, B., P. Cloutier, D. Hunting, M. Huels, and L. Sanche. 2000. Resonant formation of DNA strand breaks by low-energy (3–20 eV) electrons. *Science* 287 (5458): 1658–1660. https://doi.org/10.1126/science.287.5458.1658.

10. Sanche, L. 2009. Biological chemistry: Beyond radical thinking. *Nature* 461 (7262): 358–359. https://doi.org/10.1038/461358a.

11. Baccarelli, I., I. Bald, F.A. Gianturco, E. Illenberger, and J. Kopyra. 2011. Electron-induced damage of DNA and its components: Experiments and theoretical models. *Physics Reports* 508 (1–2): 1–44. https://doi.org/10.1016/j.physrep.2011.06.004.

12. Smyth, M., and J. Kohanoff. 2011. Excess electron localization in solvated DNA bases. *Physical Review Letters* 106 (23): 238108. https://doi.org/10.1103/PhysRevLett.106.238108.

13. Lisowski, M., P. Loukakos, U. Bovensiepen, J. Stähler, C. Gahl, and M. Wolf. 2004. Ultra-fast dynamics of electron thermalization, cooling and transport effects in Ru(001). *Applied Physics A: Materials Science & Processing* 78 (2): 165–176. https://doi.org/10.1007/s00339-003-2301-7.

14. Pellegrini, C., A. Marinelli, and S. Reiche. 2016. The physics of X-ray free-electron lasers. *Reviews of Modern Physics* 88 (1): 015006. https://doi.org/10.1103/RevModPhys.88.015006.

15. Kaganov, M.I., I.M. Lifshitz, and L.V. Tanatarov. 1957. Relaxation between electrons and the crystalline lattice. *Soviet Physics JETP* 4: 173–178.

16. Anisimov, S.I., B.L. Kapeliovich, and T.L. Perel'man. 1975. Electron emission from metal surfaces exposed to ultrashort laser pulses. *Journal of Experimental and Theoretical Physics* 39: 375–377.

17. Cho, B.I., K. Engelhorn, A.A. Correa, T. Ogitsu, C.P. Weber, H.J. Lee, J. Feng, P.A. Ni, Y. Ping, A.J. Nelson, D. Prendergast, R.W. Lee, R.W. Falcone, and P.A. Heimann. 2011. Electronic structure of warm dense copper studied by ultrafast X-ray absorption spectroscopy. *Physical Review Letters* 106 (16): 167601. https://doi.org/10.1103/PhysRevLett.106.167601.

18. Frenkel, D., and B. Smit. 2002. *Understanding molecular simulation*. Academic Press.

19. Duffy, D.M., and A.M. Rutherford. 2007. Including the effects of electronic stopping and electron-ion interactions in radiation damage simulations. *Journal of Physics: Condensed Matter* 19 (1): 016207. https://doi.org/10.1088/0953-8984/19/1/016207.

20. Köuppel, H., W. Domcke, and L.S. Cederbaum. 2007. Multimode molecular dynamics beyond the born-oppenheimer approximation. *Advances in Chemical Physics* 57: 59–246. ISBN 0065-2385. https://doi.org/10.1002/9780470142813.ch2.

21. Tavernelli, I. 2015. Nonadiabatic molecular dynamics simulations: Synergies between theory and experiments. *Accounts of Chemical Research* 48 (3): 792–800. https://doi.org/10.1021/ar500357y.

22. Horsfield, A.P., D.R. Bowler, H. Ness, C.G. Sánchez, T.N. Todorov, and A.J. Fisher. 2006. The transfer of energy between electrons and ions in solids. *Reports on Progress in Physics* 69 (4): 1195–1234. https://doi.org/10.1088/0034-4885/69/4/R05.

23. Horsfield, A.P., D.R. Bowler, A.J. Fisher, T.N. Todorov, and M.J. Montgomery. 2004. Power dissipation in nanoscale conductors: classical, semi-classical and quantum dynamics. *Journal of Physics: Condensed Matter* 16 (21): 3609–3622. https://doi.org/10.1088/0953-8984/16/21/010.

24. Horsfield, A.P., D.R. Bowler, A.J. Fisher, T.N. Todorov, and C.G. Sánchez. 2004. Beyond ehrenfest: Correlated non-adiabatic molecular dynamics. *Journal of Physics: Condensed Matter* 16 (46): 8251–8266. https://doi.org/10.1088/0953-8984/16/46/012.

25. Horsfield, A.P., D.R. Bowler, A.J. Fisher, T.N. Todorov, and C.G. Sanchez. 2005. Correlated electron-ion dynamics: The excitation of atomic motion by energetic electrons. *Journal of Physics: Condensed Matter* 17 (30): 4793–4812. https://doi.org/10.1088/0953-8984/17/30/006.

26. Bonča, J., and S.A. Trugman. 1995. Effect of inelastic processes on tunneling. *Physical Review Letters* 75 (13): 2566–2569. https://doi.org/10.1103/PhysRevLett.75.2566.

27. Rizzi, V., T.N. Todorov, J.J. Kohanoff, and A.A. Correa. 2016. Electron–phonon thermalization in a scalable method for real-time quantum dynamics. *Physical Review B* 93 (2): 024306. https://doi.org/10.1103/PhysRevB.93.024306.

28. Rizzi, V., T.N. Todorov, and J.J. Kohanoff. 2017. Inelastic electron injection in a water chain. *Scientific Reports* 7: 45410. https://doi.org/10.1038/srep45410.

# Chapter 2
# Physical Motivation

In this chapter we present a selection of physical problems where the method in this thesis with its real time treatment of the electron–phonon dynamics can offer significant insights. The problems put this work in a physical frame and, above all, provide a motivation for its findings. First, we cover the general problem of radiation damage, picking results in the damage to biological systems and metals. Second, we explore the ultrashort heating of metals with a laser and the subsequent warm dense matter dynamics.

## 2.1 Radiation Damage

Radiation damage is a topic of the utmost importance. Its range of applications is broad and spans many fields of study: from nuclear reactors design to shielding cosmic rays and improving medical treatment and imaging. For instance, the need for economically efficient energy sources with a low carbon footprint has brought nuclear power back to the world's attention. Materials in nuclear power applications are irradiated by high energy products of fission or fusion. This irradiation has two effects: it generates heat to power the nuclear plant and it damages materials, degrading their mechanical properties [1].

The damage can be due to the formation of defects in their lattices, such as vacancies and interstitials. These defects lead to macroscopic phenomena like swelling (Fig. 2.1), where gas fills the voids and makes the material expand. In general, radiation causes amorphisation of the target. The resulting change in mechanical properties can be significant and can lead to dramatic failures. Therefore it is important to discover materials that can sustain intense fluxes of radiation for extended periods of time. Ideally new materials should be able to dissipate heat quickly over a large area or to heal themselves. A better understanding of the mechanisms driving radiation damage would offer directions to this search.

© Springer International Publishing AG, part of Springer Nature 2018
V. Rizzi, *Real-Time Quantum Dynamics of Electron-Phonon Systems*,
Springer Theses, https://doi.org/10.1007/978-3-319-96280-1_2

**Fig. 2.1** Stainless rods
before (left) and after (right)
irradiation [2]. The irradiated
sample shows swelling
caused by an accumulation
of gas in the lattice
interstitials

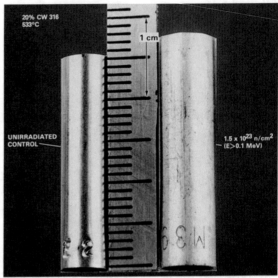

Reprinted from [L.K. Mansur, Journal of Nuclear Materials,
vol. 216, pp. 97-123, 1994], Copyright 1994,
with permission from Elsevier

New materials are strongly needed by the nuclear industry, as more resistant containment vessels for nuclear reactors would reduce upkeep and maintenance costs. Even when the actual generation of fission power plants will be phased out, such materials will be relevant for long term storage of existing nuclear waste. If nuclear fusion power plants will ever become a viable energy source, the requirement for radiation sturdy materials will be even more pressing: neutrons emerging from a deuterium-tritium reaction have an energy of 14 MeV, at least an order of magnitude more than the products of nuclear fission [3].

The need for radiation resistant materials applies to the space industry as well. The intensity of cosmic radiation in space is much higher than on the Earth's surface where the atmosphere and the planet's magnetic field deflect a large part of the radiation. Besides, one of the main factors hampering the human dream of outer space exploration is the large radiation dose that would affect the astronauts during years-long journeys.

Another crucial application for radiation damage studies is medical. The details of how radiation therapy kills cells and the role of radiation induced electrons are not yet entirely understood [4–6]. Grasping the microscopic details of the effect of radiation on biological objects, like DNA, would prove to be an invaluable source of information for cancer therapy.

### 2.1.1 Radiation Damage in Metals

A typical radiation event in a metal forms a radiation cascade. A cascade begins with a high energy particle approaching and hitting a target material (Fig. 2.2a). The series of events that follow can be split up in stages and is known as a *collision cascade*: a chain of displacement events in the target material with energy spread and dissipation. The first target atom that is struck (Fig. 2.2b) is called *primary knock-on atom* (PKA) and its energy spectrum depends both on the colliding particle and the target material. For example, in fusion reactors a 14 MeV neutron produces iron PKAs up to 1 MeV energy,[1] half of them with an energy over 10 keV.

The critical factor for the next phases is the cross section between the PKA and the surrounding target atoms. The cross section strongly depends on the energy of the PKA: when the PKA has a high energy (i.e. above 100 keV), its interaction cross section with the nuclei of the target will be low, so it will travel long distances before new collisions. This *channelling* phase (Fig. 2.2c) will determine the penetration depth of the PKA and the extent of the damage to the material. The electrons of the target atoms play an important role to the PKA losing energy (Fig. 2.2d). An amorphous solid will have a shallower penetration depth than a crystal [7].

With the channelling PKA slowing down, its impact cross-section with the surrounding atoms increases until the PKA impacts another atom (Fig. 2.2e). The two particles will either keep channelling through the target (a *sub-cascade branching*, Fig. 2.2f) if their energy is relatively high ($\gtrsim$10 keV [7]),[2] or they will trigger a new stage of the cascade, the *displacement phase*.

This new phase is characterized by strong interactions between the projectiles and the target atoms and it features a large number of collisions. In picoseconds, most atoms in a region of 10–100 nm are displaced, giving rise to a *displacement spike* (Fig. 2.2g). *Replacement collision sequences* (Fig. 2.2h) are common and consist of atoms being displaced sequentially along crystal axes directions, as in colliding lined up billiard balls.

After a *relaxation phase*, when the energy is spread throughout the displacement region and defects appear (Fig. 2.2i), there is a *cooling phase* which can be hundreds of picoseconds long. Heat is dissipated, while interstitials and vacancies tend to recombine (Fig. 2.2j) until a final equilibrium configuration of the system is reached (Fig. 2.2k).

---

[1] One could wonder if it is appropriate to take special relativity into account at this scale of energies. The rest energy for an iron atom is $\approx$50 GeV, several orders of magnitude above the energies considered in the collisions, so special relativity can be safely ignored in this case. By contrast, for cosmic-ray cascades, relativity plays a crucial role since the projectiles are usually light and extremely fast.

[2] It is intuitive that the higher the initial energy of the PKA, the larger the number of sub-cascades expected. A recent article [8] studies high energy iron cascades (up to 0.5 MeV). Contrary to intuition and common belief, the authors reported a reduced cascade branching compared to lower energy cascades and a rather continuous distribution of damage.

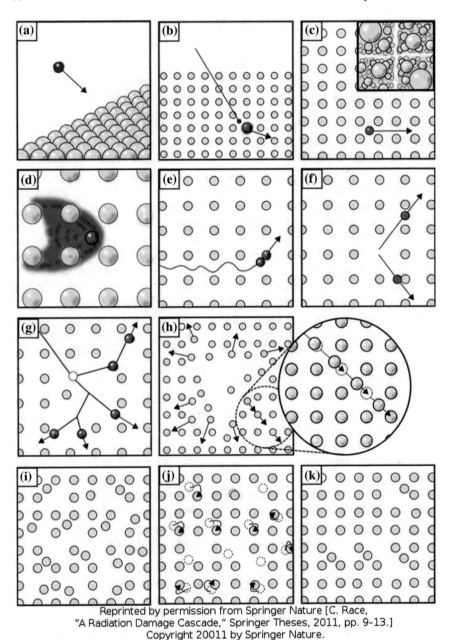

**Fig. 2.2** Image that shows the stages of a radiation cascade from [7]

### 2.1.1.1 Accounting for the Energy Loss, the Role of Electrons

The schematic in last section describes a radiation event from the point of view of target atoms. In some stages of the cascade, electrons play a significant part in determining the final extent of radiation damage.

In the channelling phase, the interaction between the high speed PKA and the atoms of the target is small. Experiments reveal tracks of damage on the target around the path of channelling atoms. Here electrons play the key role of moving energy from the fast projectile to the surrounding environment. Models that try to predict the rate of energy loss can be found in Chap. 3. Several schemes have been proposed to explain this dissipation mechanism [7]. One is called *Coulomb explosion model* and suggests that excited electrons are ballistically ejected from the track region. The depletion of charge along the track forms a positively charged region that may lead to a shock wave. Another model describes the transfer of heat from the excited electrons in the track region to the surrounding ions through the electron–phonon interaction, the *thermal spike model*.

The incoming energy is initially concentrated in the PKA. During the channelling phase and the subsequent displacement phase, most of the energy is in the projectile and in the atomic system. The role of the electrons along the atomic tracks is to influence the atomic motion, by dampening or heating atoms and causing a net flow of energy from or into the atoms. Recent ab-initio simulations based on Ehrenfest dynamics have been employed to estimate this energy exchange between atoms and electrons [3, 9–11]. Electronic excitations were found to alter inter-atomic forces as electrons lose their ability to provide chemical bonding as they become excited (Fig. 2.3).

In the cooling phase, the excited electrons work as a heat bath for the atoms. Because of their high thermal conductivity, electrons can increase the speed of cooling in a metal providing a channel for enhanced energy transport away from the hot ionic areas. They can also influence the production and healing of defects since they play an active role in the potential felt by the atoms. To go beyond models where either the electrons or the atoms are a heat bath for the other subsystem is a very challenging problem. The non-adiabatic dynamics of secondary electrons is not yet completely understood. They can lose energy to vibrations via spontaneous phonon emission, but this process is not captured by Ehrenfest dynamics. The possibility to study in real time the mutual energy exchange of electrons and atoms could offer precious insights on the microscopic detail of radiation damage.

## 2.1.2 Radiation Damage in Biological Systems

X-rays or high-energy particles hitting cells can ionise molecules on their path, mainly water, the most common molecule in a biological system. The secondary electrons emitted interact with other electrons and phonons and lose energy to them, until they reach the DNA. Depending on their energy spectrum, they can cause

**Fig. 2.3** Illustration of the electronic wake (blue surfaces) generated by an energetic proton (red sphere) traveling in an aluminum crystal (yellow spheres). The resulting change in electronic density is responsible for modification of chemical bonds between the atoms and consequently for a change in their interactions [9]

damage to the cell, possibly leading to its death. While the initial conditions at the start of a radiation event and the damage afterwards are generally well understood, today there is no clear microscopic understanding of the sequence of processes that leads to cell death [6]. The study of these processes is tied to the dynamics of electrons, from their excitation by the external radiation to their eventual effect on DNA. The electron–phonon interaction plays a crucial role in how the electrons' propagation unfolds. About $\approx 50 \times 10^3$ secondary electrons are emitted for every MeV of incoming energy [12, 13] and most of them are low energy electrons (LEEs), with an energy distribution peaking below 10 eV [14] (Fig. 2.4).

The phase with the largest impact on the evolution of an irradiated cell is the fast and ultrafast one that follows the radiation event [6] on a femtosecond to picosecond scale and about two thirds of the cellular damage is caused by secondary species [15]. The processes triggered at these timescales can, at later times, induce fractures in the structure of DNA: either reparable single strand breaks or lethal double strand breaks [16].

One could imagine that DNA damage increases with the incoming electronic energy and that there is a threshold energy below which the electrons cannot ionize DNA or trigger any harmful process. This belief was questioned by [4, 15], where it was discovered that LEEs with energies between 3 and 20 eV can damage DNA considerably and their damaging power does not constantly increase with their energy. These studies triggered an intense effort into understanding the interaction mecha-

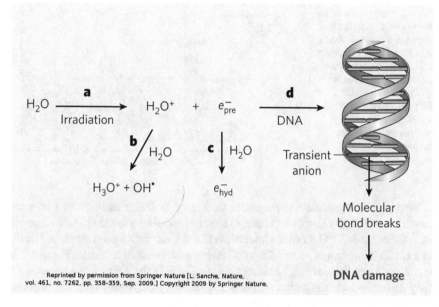

**Fig. 2.4** A possible mechanism of radiation damage in DNA [12]. Irradiated water ionizes and the short lived pre-hydrated secondary electrons can react with DNA bases, forming transient anions and leading to bond breaking and DNA damage.

nisms of LEEs with DNA [6, 13, 17–19]. The capability of LEEs of causing single and double strand breaks to DNA is now widely accepted [6, 13], but the mechanism that leads to the DNA damage is not yet completely understood. The study of LEEs' dynamics and the interactions with the surrounding cellular environment can offer a better understanding of their interplay with DNA and, ultimately, lead to improvements in radiotherapy. For example, secondary electrons can cause different levels of heating to the water in cells depending on their energy spectrum. The assessment of electron-water energy transfers can improve the understanding of the subsequent damage.

There has been an intense experimental effort on the damage of LEEs on DNA, mostly in a gas or condensed phase, but such experiments do not show how electrons can damage DNA indirectly, through the environment [13]. Experiments involving LEEs in water are quite challenging. A recent study [20] revealed that the presence of an enviroment of water and oxygen leads to a significant increase in the formation of double strand breaks in DNA, almost seven times higher than in vacuum. It was theorized that dissociative states of a water-DNA complex can contribute to the DNA damage significalntly [21] and that a key role is played by slow electrons scattering inelastically and exciting shape resonances. The double strand breaks that would form in connection with the presence of water are difficult to repair because of the close proximity of the damage [13].

**Fig. 2.5** Schematics of the equilibration process of electrons in a metal after a laser excitation [27]. Before and well after the excitation, the electronic system shows FD distributions with different temperatures, while just after excitation it presents a strongly non thermal distribution

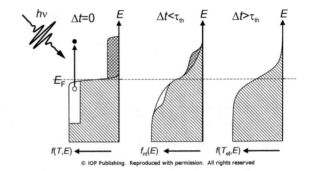

First principles molecular dynamics studies of DNA bases immersed in water show that an initially delocalized excess electron localizes around the nucleobase in a few femtoseconds [5]. Further simulations [22, 23] emphasize the role played by the water in this context and show a different behaviour depending on the nucleobase. The starting point of these adiabatic simulations is a LEE in the vicinity of a nucleobase. They do not include the earlier non-adiabatic electronic dynamics that could reveal the dependence of the damage on electronic energy.

A key mechanism leading to nucleobase damage is dissociative electron attachment (DEA) [24]. The impact of DEA on two different nuclebases in the presence of water was investigated in [25], where it was found that the presence of the water strongly enhances the DEA cross section. Amino acids were found to have a protective role for nucleobases in the presence of excess electrons [26].

At the heart of these studies lies the need to unravel the actions of a single track of radiation on a cell [15]. The history of an electron with all its interactions can reveal the microscopic details of where the radiation impact is most severe. For example, at environmental levels of exposure, the cells in a human body only experience electron tracks every few months [15]. A knowledge of how *individual* electron tracks inelastically interact with the aqueous biological environment and DNA can lead to more refined models for calculating human risk at standard exposure levels.

## 2.2 Ultrashort Laser Heating of Metals

The irradiation of a metal with an ultrashort laser pulse leads to the energy being absorbed by the electron system, which gets driven out-of-equilibrium (OOE) on a femtosecond time scale. The electronic heat capacity in metals is typically orders of magnitude smaller than the one of the lattice, thus a sub-picosecond laser pulse can selectively excite electrons to temperatures in the range of thousands of K, while keeping the underlying lattice cold [28, 29] (Fig. 2.5).

The relaxation of these OOE electrons to a thermal Fermi-Dirac (FD) distribution involves electron–electron (e–e) scattering or the usually slower electron–phonon

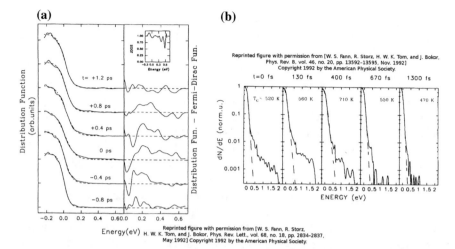

**Fig. 2.6** In **a**, photoelectron energy distributions (solid line) and Fermi-Dirac fit (dashed line) for various times delay [28]. The FD fits correspond to electron temperatures of 380, 483, 625, 582, 508, and 490 K for time delays of −0.8, −0.4, 0.0, +0.4, +0.8, and +1.2 ps respectively. On the right, a 10x blow up of the curves. In **b**, electron energy distributions after laser excitation at different times [35]. The dashed line is the best FD fit at the temperature indicated in the figure

(e–p) interaction. Anyhow, relaxation times are strongly dependent on the energy of the OOE electrons and on the temperature of the phonon [30]. After a new thermodynamical equilibrium state is reached in the electronic system, energy can flow from the hot electrons to the phonons. This is the basis of the well known two-temperature model [31] (see Chap. 3), where it is assumed that electrons immediately thermalize, before they start to interact with the phonons [32, 33].

This equilibration is a complicated many-body problem and the assumption of an instantaneous electronic excitation ignores important aspects of the process. An OOE gas couples differently to the phononic system compared to a thermalized one. Moreover, the thermalization time and the e–p coupling strength are both strongly dependent on material properties and excitation type [34].

Time-resolved photoemission spectroscopy experiments can track electronic dynamics by measuring electron energy distributions at different times. Early photoemission experiments were performed in 1992 on a gold film [28, 35], where the energy distribution was compared to a FD fit. In Fig. 2.6a we can see that the electronic population displays large and systematic deviations from a FD distribution, notably for time delays at about 0 ps in the energy range 0–0.5 eV. The laser beam excites electrons away from a thermal distribution and their equilibration occurs over a finite time. The hot tail in the energy distributions at intermediate times after the excitation is a sign of non-thermalized excited electrons and is the reason why a FD fit would not be physically appropriate.

In Fig. 2.6b an analogous experiment is performed with an improved time resolution. Without any energy relaxation mechanism, a laser would cause a constant excitation in the energy range from 0 to the pump photon energy 1.84 eV. Fermi liquid theory provides a scaling of $(E - E_F)^{-2}$ for the relaxation time of an electron with energy $E$ under e–e interaction, so the highest excited electrons take the shortest time to relax. At 130 fs the excited electron population is almost flat, while at 400 fs most of the highly excited electrons are thermalized and the best-fit FD temperature reaches its maximum at 710 K. By 670 fs the electrons are nearly at equilibrium at a temperature of 550 K. Evidently the electrons do lose energy to the lattice subsystem *before* they reach thermalization. The authors assume an average electron–phonon collision time of 30 fs, so in a ps time window several collisions can occur, producing a substantial energy exchange between electrons and phonons. This shows that the electron distribution can be nonthermal on the e–p timescale.

Several studies have been conducted on this topic over the years. A work on Ru(001) [36] shows clear deviations from a thermal FD distribution within the first 500 fs of the dynamics. In Fig. 2.7a, the photoelectron energy distributions before and after excitation are compared and fitted with FD distributions. The spectrum before irradiation is thermalized at 100 K, whereas, after irradiation, only 80% of the electrons can be accounted by a FD distribution corresponding to 225 K. A significant portion of the electrons are non-thermal and populates states up to 1.2 eV above the Fermi energy. The dynamics of the photoelectron distributions between 0 and 500 fs is displayed in Fig. 2.7b for three different laser fluences. For higher fluences, the relative number of non-thermal electrons increases as well as the temperature of the thermal distributions. The maximum population of non-thermal electrons is reached at a delay of 100 fs, while at 500 fs the electrons are essentially thermalized. The dynamics of the energy exchanges can be rationalized by a competition between energy transfer from electrons to phonons and ballistic transport. While the thermal electrons tend to localize energy on the surface of the material through e–p coupling, the non-thermal ones tend to carry heat away in the bulk through ballistic transport. The authors propose an extended heat-bath model that goes beyond the usual two-temperature model (2TM) and treats separately thermal and non-thermal electrons.

A photoemission experiment on the lanthanide ferromagnetic metal Gd(001) [27, 37] obtains similar results, albeit with a less pronounced effect. In Fig. 2.7c, the spectrum at time delay zero displays a kink that signals the presence of non-thermal electrons. For delays larger than 300 fs, the kink disappears and the spectra reasonably agree with a FD distribution.

In [38], the authors perform time resolved electron spectroscopy on cuprate superconductors and measure energy-resolved population lifetimes $\tau_p(\epsilon)$. In Fig. 2.8c–e, the population dynamics at $T = 20$ K shows a clear change in behaviour near 60 meV. Below that energy, the excited population grows until 0.5 ps and then slowly decays over a few ps. Above that energy, the population reaches its maximum at time 0 and decays fast over a few hundreds of fs. For $T = 120$ K in Fig. 2.8g–i, the energy dependence of the dynamics is less evident. The comparison of the population decay rates with the single-particle lifetimes shows discrepancies of 1–2 orders of magnitude, making clear that the two lifetimes are distinct. To rationalize the dynamics, the authors propose a population model based on collision integrals that include different electron scattering channels. Although offering a better qualitative

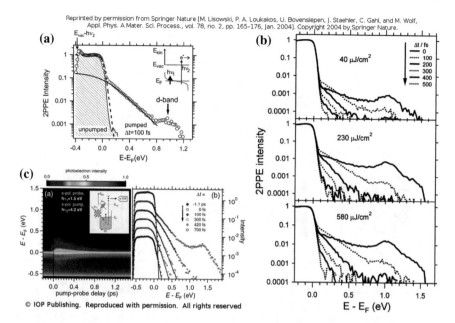

**Fig. 2.7** In **a–b**, photoelectron energy distributions for Ru(001) [36], obtained from time-resolved two-photon photo-electron (2PPE) spectroscopy experiments. In **a**, there is a comparison of the unpumped and pumped (after 100 fs) spectra with the dashed line a FD distribution at temperature 225 K and the thick solid line an approximation of the non thermal part of the spectrum. In **b**, different time delays and different fluences are shown. In **c**, photoemission intensity in Gd(001) as a function of energy and time delay (left) and energy spectra for different time delays (right) [27]

understanding of the physics at play, the kinetic model is not enough to capture the nontrivial population decay. They wish for a more complete model to interpret and compare their injected electrons energy dependent dynamics.

Today the dynamical interplay of phonons with optically excited electrons has sparkled interest in a wide range of materials: from warm dense matter [39, 40] to high temperature superconductors [38, 41, 42]. It is understood that, at early times after photoexcitation, non-thermal electrons are very relevant in the dynamics and a 2TM description is valid only after complete electron thermalization. There is a need to go beyond simple models to describe OOE electronic dynamics interacting with phonons. Knowledge about the time dependent non-equilibrium electron distribution and its thermalization dynamics can provide vital information for surface femtochemistry [36].

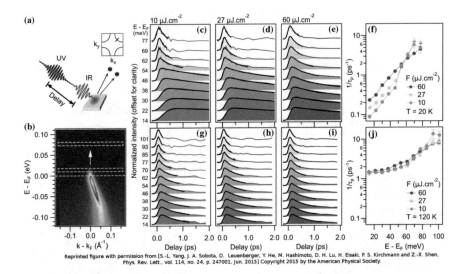

**Fig. 2.8** Energy-resolved population decay in a photoemission experiment on a cuprate superconductor [38]. In **a**, a sketch of the pump-probe experiment. In **b**, nodal cut from a 6 eV photoemission at $T = 20$ K. In **c–e**, population dynamics for different photoemission fluences at $T = 20$ K and in **g–i** the same at $T = 120$ K. In **f** and **j**, fits of the population decay rates at different temperatures and fluences

# References

1. Zarkadoula, E., R. Devanathan, W.J. Weber, M.A. Seaton, I.T. Todorov, K. Nordlund, M.T. Dove, and K. Trachenko. 2014. High-energy radiation damage in zirconia: Modeling results. *Journal of Applied Physics* 115 (8): 083507. https://doi.org/10.1063/1.4866989.

2. Mansur, L. 1994. Theory and experimental background on dimensional changes in irradiated alloys. *Journal of Nuclear Materials* 216: 97–123. https://doi.org/10.1016/0022-3115(94)90009-4.

3. Race, C.P., D.R. Mason, M.W. Finnis, W.M.C. Foulkes, A.P. Horsfield, and A.P. Sutton. 2010. The treatment of electronic excitations in atomistic models of radiation damage in metals. *Reports on Progress in Physics* 73 (11): 116501. https://doi.org/10.1088/0034-4885/73/11/116501.

4. Boudaïffa, B., P. Cloutier, D. Hunting, M. Huels, and L. Sanche. 2000. Resonant formation of DNA strand breaks by low-energy (3–20 eV) electrons. *Science* 287 (5458): 1658–1660. https://doi.org/10.1126/science.287.5458.1658.

5. Smyth, M., and J. Kohanoff. 2011. Excess electron localization in solvated DNA bases. *Physical Review Letters* 106 (23): 238108. https://doi.org/10.1103/PhysRevLett.106.238108.

6. Baccarelli, I., I. Bald, F.A. Gianturco, E. Illenberger, and J. Kopyra. 2011. Electron-induced damage of DNA and its components: Experiments and theoretical models. *Physics Reports* 508 (1–2): 1–44. https://doi.org/10.1016/j.physrep.2011.06.004.

7. Race, C. 2011. *A radiation damage cascade*. In *The modelling of radiation damage in metals using ehrenfest dynamics*. Springer Theses, Springer. https://doi.org/10.1007/978-3-642-15439-3_2.

8. Zarkadoula, E., S.L. Daraszewicz, D.M. Duffy, M.A. Seaton, I.T. Todorov, K. Nordlund, M.T. Dove, and K. Trachenko. 2013. The nature of high-energy radiation damage in iron. *Journal of*

*Physics. Condensed Matter: An Institute of Physics Journal* 25(12): 125402. https://doi.org/ 10.1088/0953-8984/25/12/125402.

9. Correa, A.A., J. Kohanoff, E. Artacho, D. Sánchez-Portal, and A. Caro. 2012. Nonadiabatic forces in ion-solid interactions: The initial stages of radiation damage. *Physical Review Letters* 108 (21): 213201. https://doi.org/10.1103/PhysRevLett.108.213201.

10. Mason, D.R., J. le Page, C.P. Race, W.M.C. Foulkes, M.W. Finnis, and A.P. Sutton. 2007. Electronic damping of atomic dynamics in irradiation damage of metals. *Journal of Physics: Condensed Matter* 19 (43): 436209. https://doi.org/10.1088/0953-8984/19/43/436209.

11. Zeb, M.A., J. Kohanoff, D. Sánchez-Portal, A. Arnau, J.I. Juaristi, and E. Artacho. 2012. Electronic stopping power in gold: The role of d electrons and the H/He anomaly. *Physical Review Letters* 108 (22): 225504. https://doi.org/10.1103/PhysRevLett.108.225504.

12. Sanche, L. 2009. Biological chemistry: Beyond radical thinking. *Nature* 461 (7262): 358–359. https://doi.org/10.1038/461358a.

13. Alizadeh, E., T.M. Orlando, and L. Sanche. 2015. Biomolecular damage induced by ionizing radiation: The direct and indirect effects of low-energy electrons on DNA. *Annual Review of Physical Chemistry* 66: 379–98. https://doi.org/10.1146/annurev-physchem-040513-103605.

14. Pimblott, S.M., and J.A. LaVerne. 2007. Production of low-energy electrons by ionizing radiation. *Radiation Physics and Chemistry* 76 (8–9): 1244–1247. https://doi.org/10.1016/ j.radphyschem.2007.02.012.

15. Michael, B.D. 2000. A sting in the tail of electron tracks. *Science* 287 (5458): 1603–1604. https://doi.org/10.1126/science.287.5458.1603.

16. Jeggo, P.A., and M. Löbrich. 2007. DNA double-strand breaks: Their cellular and clinical impact? *Oncogene* 26 (56): 7717–9. https://doi.org/10.1038/sj.onc.1210868.

17. Martin, F., P. Burrow, Z. Cai, P. Cloutier, D. Hunting, and L. Sanche. 2004. DNA strand breaks induced by 04 eV electrons: The role of shape resonances. *Physical Review Letters* 93 (6): 068101. https://doi.org/10.1103/PhysRevLett.93.068101.

18. Cho, W., M. Michaud, and L. Sanche. 2004. Vibrational and electronic excitations of $H_2O$ on thymine films induced by low-energy electrons. *The Journal of Chemical Physics* 121 (22): 11289. https://doi.org/10.1063/1.1814057.

19. Simons, J. 2006. How do low-energy (0.1–2 eV) electrons cause DNA strand breaks? *Accounts of Chemical Research* 39 (10): 772–779. https://doi.org/10.1021/ar0680769.

20. Alizadeh, E., and L. Sanche. 2013. Role of humidity and oxygen level on damage to DNA induced by soft X-rays and low-energy electrons. *The Journal of Physical Chemistry C* 117 (43): 22 445–22 453. https://doi.org/10.1021/jp403350j.

21. Orlando, T.M., D. Oh, Y. Chen, and A.B. Aleksandrov. 2008. Low-energy electron diffraction and induced damage in hydrated DNA. *The Journal of Chemical Physics* 128 (19): 195102. https://doi.org/10.1063/1.2907722.

22. Smyth, M., and J. Kohanoff. 2012. Excess electron interactions with solvated DNA nucleotides: Strand breaks possible at room temperature. *Journal of the American Chemical Society* 134 (22): 9122–9125. https://doi.org/10.1021/ja303776r.

23. McAllister, M., M. Smyth, B. Gu, G.A. Tribello, and J. Kohanoff. 2015. Understanding the interaction between low-energy electrons and DNA nucleotides in aqueous solution. *The Journal of Physical Chemistry Letters* 6 (15): 3091–3097. https://doi.org/10.1021/acs.jpclett.5b01011.

24. Haxton, D.J., Z. Zhang, H.-D. Meyer, T.N. Rescigno, and C.W. McCurdy. 2004. Dynamics of dissociative attachment of electrons to water through the 2B1 metastable state of the anion. *Physical Review A* 69 (6): 062714. https://doi.org/10.1103/PhysRevA.69.062714.

25. Smyth, M., J. Kohanoff, and I.I. Fabrikant. 2014. Electron-induced hydrogen loss in uracil in a water cluster environment. *The Journal of Chemical Physics* 140 (18): 184313. https://doi.org/10.1063/1.4874841.

26. Gu, B., M. Smyth, and J. Kohanoff. 2014. Protection of DNA against low-energy electrons by amino acids: A first-principles molecular dynamics study. *Physical Chemistry Chemical Physics* 16 (44): 24 350–24 358. https://doi.org/10.1039/C4CP03906H.

27. Bovensiepen, U. 2007. Coherent and incoherent excitations of the Gd(0001) surface on ultrafast timescales. *Journal of Physics: Condensed Matter* 19 (8): 083201. https://doi.org/10.1088/ 0953-8984/19/8/083201.

28. Fann, W.S., R. Storz, H.W.K. Tom, and J. Bokor. 1992. Direct measurement of nonequilibrium electron-energy distributions in subpicosecond laser-heated gold films. *Physical Review Letters* 68 (18): 2834–2837. https://doi.org/10.1103/PhysRevLett.68.2834.
29. Del Fatti, N., C. Voisin, M. Achermann, S. Tzortzakis, D. Christofilos, and F. Vallée. 2000. Nonequilibrium electron dynamics in noble metals. *Physical Review B* 61 (24): 16 956–16 966. https://doi.org/10.1103/PhysRevB.61.16956.
30. Ziman, J. M. 1972. Transport properties. In *Principles of the theory of solids*, 211–254. Cambridge: Cambridge University Press.
31. Anisimov, S.I., B.L. Kapeliovich, and T.L. Perel'man. 1975. Electron emission from metal surfaces exposed to ultrashort laser pulses. *Journal of Experimental and Theoretical Physics* 39: 375–377.
32. Kaganov, M.I., I.M. Lifshitz, and L.V. Tanatarov. 1957. Relaxation between electrons and the Crystalline Lattice. *Soviet Physics JETP* 4: 173–178.
33. Tzou, D. 2014. Ultrafast pulse-laser heating on metal films. In *Macro- to microscale heat transfer*, 193–229. Chichester, UK: Wiley. https://doi.org/10.1002/9781118818275.ch5.
34. Mueller, B.Y., and B. Rethfeld. 2013. Relaxation dynamics in laser-excited metals under nonequilibrium conditions. *Physical Review B* 87 (3): 035139. https://doi.org/10.1103/PhysRevB.87.035139.
35. Fann, W.S., R. Storz, H.W.K. Tom, and J. Bokor. 1992. Electron thermalization in gold. *Physical Review B* 46 (20): 13 592–13 595. https://doi.org/10.1103/PhysRevB.46.13592.
36. Lisowski, M., P. Loukakos, U. Bovensiepen, J. Stähler, C. Gahl, and M. Wolf. 2004. Ultrafast dynamics of electron thermalization, cooling and transport effects in Ru(001). *Applied Physics A: Materials Science & Processing* 78 (2): 165–176. https://doi.org/10.1007/s00339-003-2301-7.
37. Bovensiepen, U., and P. Kirchmann. 2012. Elementary relaxation processes investigated by femtosecond photoelectron spectroscopy of two-dimensional materials. *Laser & Photonics Reviews* 6 (5): 589–606. https://doi.org/10.1002/lpor.201000035.
38. Yang, S.-L., J. Sobota, D. Leuenberger, Y. He, M. Hashimoto, D. Lu, H. Eisaki, P. Kirchmann, and Z.-X. Shen. 2015. Inequivalence of single-particle and population lifetimes in a cuprate superconductor. *Physical Review Letters* 114 (24): 247001. https://doi.org/10.1103/PhysRevLett.114.247001.
39. Cho, B.I., K. Engelhorn, A.A. Correa, T. Ogitsu, C.P. Weber, H.J. Lee, J. Feng, P.A. Ni, Y. Ping, A.J. Nelson, D. Prendergast, R.W. Lee, R.W. Falcone, and P.A. Heimann. 2011. Electronic structure of warm dense copper studied by ultrafast X-ray absorption spectroscopy. *Physical Review Letters* 106 (16): 167601. https://doi.org/10.1103/PhysRevLett.106.167601.
40. Ogitsu, T., Y. Ping, A. Correa, B.I. Cho, P. Heimann, E. Schwegler, J. Cao, and G.W. Collins. 2012. Ballistic electron transport in non-equilibrium warm dense gold. *High Energy Density Physics* 8 (3): 303–306. https://doi.org/10.1016/j.hedp.2012.01.002.
41. Perfetti, L., P.A. Loukakos, M. Lisowski, U. Bovensiepen, H. Eisaki, and M. Wolf. 2007. Ultrafast electron relaxation in superconducting Bi(2)Sr(2)CaCu(2)O(8+delta) by time-resolved photoelectron spectroscopy. *Physical Review Letters* 99 (19): 197001. https://doi.org/10.1103/PhysRevLett.99.197001.
42. Avigo, I., R. Cortés, L. Rettig, S. Thirupathaiah, H.S. Jeevan, P. Gegenwart, T. Wolf, M. Ligges, M. Wolf, J. Fink, and U. Bovensiepen. 2013. Coherent excitations and electron–phonon coupling in Ba/EuFe2As2 compounds investigated by femtosecond time- and angle-resolved photoemission spectroscopy. *Journal of Physics: Condensed Matter* 25 (9): 094003. https://doi.org/10.1088/0953-8984/25/9/094003.

# Chapter 3
# Simulating Electrons and Phonons: Effective Temperature Methods

Heat transport can be seen microscopically as a sequence of collisions between energy carriers. The concepts of mean free path and mean free time play a key role in heat transport problems, the former being the average distance that a carrier covers between two collisions and the latter the average time, taken over a large number of collision events.

Macroscopic models require very large length and time-scales, so that many thousands of collisions occur before making an observation on the heat transport process. In solid state physics, electrons and phonons play a key role in thermalization. The mean free time between collisions is a quantity that depends on the electronic energy and on the phonon temperature [1]. Typically in a metal electron-electron scattering happens at a much faster timescale than electron-phonon and phonon-phonon processes.

If for macroscopic systems an average description over many collisions would be enough to capture the underlying physics of heating, in microscopic devices the averaging process would wash out essential details. The need for an in-depth characterization of out-of-equilibrium systems has led to the development of a number of heat transfer models.

In this chapter we present a selection of electron-phonon methods that follow the dynamics of the electronic and phononic subsystems introducing an effective temperature description.

© Springer International Publishing AG, part of Springer Nature 2018
V. Rizzi, *Real-Time Quantum Dynamics of Electron-Phonon Systems*,
Springer Theses, https://doi.org/10.1007/978-3-319-96280-1_3

## 3.1  An Effective Temperature Model for Radiation Cascades

An early attempt of including the electronic contribution in the dynamics of a radiation event in a metal was made in 1988 [2]. Until that seminal paper, the contribution of metal valence electrons during a radiation cascade had been neglected or demoted to an energy sink for slowing down fast incoming particles. That article introduced the Flynn-Averback model and paved the way for the successive creation of more advanced methods that included the electronic contribution in models for radiation cascades.

An energetic atom that collides with a lattice deposits its energy in a region of radius $r$ that heats up rapidly to a high temperature. The dimension of the thermal spike region increases as temperature decreases and heat is dissipated into the lattice. Roughly speaking, a 10 keV collision leaving 1 eV per atom in a thermal spike region of radius $r \approx 30$Å would reach a uniform temperature of $\approx 10^4$ K. Early MD simulations show that, for the temperature to decrease significantly below the melting point of the material, the required cooling time would have to be several ps [3]. Those calculations ignore the electronic contribution to thermalization, while the role of the electrons has in fact proven to be pivotal [2].

Suppose that in a metal the electronic mean free path (EMFP) $\lambda$ can be written as

$$\lambda = r_s \frac{T_0}{T_1} \tag{3.1}$$

where $r_s$ is the Wigner-Seitz (WS) radius of the material, $T_1$ is the lattice temperature and $T_0$ is the temperature at which $\lambda = r_s$. $T_0$ effectively measures the strength of the electron-phonon coupling: a low $T_0$ implies a strong coupling and a correspondingly large energy exchange. $T_1$ is restricted to the range $\theta_D < T_1 < T_0$ where $\theta_D$ is the Debye temperature.[1] Energy conservation among the $r^3/r_s^3$ involved atoms imposes a radius for the spike region

$$r = r_s \left( \frac{E}{3k_B T_1} \right)^{\frac{1}{3}} \tag{3.2}$$

where $E$ is the energy in the system.

For small enough $T_1$ (that are ultimately reached at the end of the cooling down period), it is clear that $\lambda > r$, so, for long enough times, the electronic system would always be out of equilibrium with the lattice. Whether the electrons can equilibrate with the lattice for shorter times depends strongly on the electron-phonon coupling

---

[1]For $T_1 > T_0$, $\lambda$ would be smaller than the WS radius and the details of the specific system would come into play significantly, while this method wants to be general and not system-specific. The choice of $T_1 > \theta_D$ is dictated by the exchange of energy between electrons and phonons. In this method it is assumed that a collision between an electron and a hot ion would always transfer an energy of $k_B \theta_D$, the maximum amount possible. This approximation is not restrictive since typical Debye temperatures are close to room temperatures, much cooler than the $T_1$ of heat spikes.

$T_0$. If initially $\lambda \ll r$, initially cold electrons would have time to heat up by scattering often with the lattice before they eventually dissipated the heat out of the hot region, while for a starting $\lambda \approx r$ the electron-lattice interaction and its heat dissipation would be less effective.

To determine the conditions for this change of behaviour, the authors of [2] employed a random walk model. The number of free paths (and thus of collisions) that an electron needs for escaping the hot region would be $r^2/\lambda^2$. With an energy transfer per collision $k_B\theta_D$, the escaping electrons would reach an effective temperature of

$$T_{eff} = \theta_D \frac{r^2}{\lambda^2}. \tag{3.3}$$

Setting $T_1 = T_{eff}$ and imposing it in the previous equations, the bounding parameters for different thermalization behaviours are a critical radius

$$r_c = \frac{r_s\theta_D Q}{3k_B T_0^2} \tag{3.4}$$

and a critical thermal spike temperature

$$T_c = \frac{9T_0^6 k_B^2}{\theta_D^3 Q^2}, \tag{3.5}$$

where $Q$ is the total energy of the spike in the lattice region. For initial $r \leq r_c$ or $T_1 \geq T_c$, the electrons would be able to reach equilibrium with the lattice system *before* escaping the spike region. In other words, a high $T_0$ makes the emfp $\lambda$ too long for an efficient heat exchange with the lattice to take place.

For example, similar materials such as Cu and Ni present respective $T_0$ of 45000 K and 15000 K. Because of the 6th power dependency of Eq. (3.5) on $T_0$, small differences in $T_0$ lead to considerably different equilibration dynamics. The $T_c$ of Cu and Ni are in fact vastly different, 200000 and 300 K, so that the hot spike electronic dissipation in Ni should be much more efficient than in Cu [2]. In Ni, electrons would moderate the effect of a thermal spike by absorbing heat from the hot spike region and then bringing it away, making the dissipation of the spike faster and colder dissipation.

## 3.2 The Two-Temperature Model (2TM)

A drawback of the Flynn-Averback model is that it doesn't consider the electronic temperature as an independent dynamic variable. Another method that can track the dynamics of both the electronic and lattice temperatures was developed as early as 1957 [4, 5], but the limited computing capabilities of the time made it impractical for numerical computation until the 1990s [6, 7]. It is called the two-temperature model

(2TM) and it has now become standard in describing electron-phonon thermalization processes in a range of system sizes. It has been used macroscopically to describe heat diffusion [8, 9] and microscopically to portray heat transfer at the atomistic level. Here we focus on the latter.

The 2TM uses the fact that the heat capacity of electrons $C_e$ is typically 1 to 2 orders of magnitude smaller than the lattice one $C_l$. This makes the respective time-scales widely different so that a heating process can be split into a fast electron excitation and a slower lattice excitation. This is the model's crucial assumption. The model considers both subsystems to be in local equilibrium at all times and their rate of heat exchange to be proportional to their temperature difference. The time dependence of the respective temperatures $T_e$ and $T_l$ is described by [10]

$$C_e \frac{\partial T_e}{\partial t} = \nabla \cdot (K \nabla T_e) - G(T_e - T_l) + P(t) \qquad (3.6)$$

$$C_l \frac{\partial T_l}{\partial t} = G(T_e - T_l) \qquad (3.7)$$

for the dynamics of an electron gas. $K$ is the temperature dependent thermal conductivity of the electrons, $G$ the electron phonon-coupling and $P$ an external time-dependent heat source. Equation (3.6) describes the energy balance of the electrons, whereas Eq. (3.7) expresses the heating of the lattice.

$G$ was determined [4] by summing over all one-phonon emission and absorption processes $G = \frac{\pi^2}{6} \frac{m_e n_e v_s^2}{\tau_e T_e}$ where $m_e$ is the electron mass, $n_e$ the electron density, $v_s$ the speed of sound and $\tau_e$ the electron mean free time. This form is valid in the limit $T_e \gg T_l$, i.e. in the early stages of the heating process, in the region of a few picoseconds from the start of the radiation cascade. $G$ is a very difficult quantity to determine. In Table 3.1 in [11], a comparison of theoretical and experimental values of $G$ for several metals shows that there is little to no agreement among a number of sources and different methods.

The difficulty in solving Eqs. (3.6) and (3.7) lies in the non trivial dependency of the parameters on the temperature. $G$ presents a strong temperature dependence at $T_e$ close to the Fermi temperature $T_F = k_B E_F$ [10] with $k_B$ being Boltzmann's constant and $E_F$ the Fermi energy. Also $C_e$ and $C_l$ can present a relevant temperature dependence.

Finnis's 2TM [6] recovers the critical temperature concept (3.5) and was used to test Flynn and Averback's hypothesis of different cooling rates in Cu and Ni [2]. To get an estimate of the electronic rate of energy acquisition from the ions, the authors consider electrons with an electronic mean free path $\lambda$ (see Eq. (3.1)), colliding with the lattice at a Fermi velocity $v_F$ and receiving an energy $k_B \theta_D$ per collision. With these assumptions, the scattering rate is $v_F / \lambda$ and the number of electrons involved in the scattering is $\approx k_B T_e D(E_F)$ where $D(E_F)$ is the electronic density of states at the Fermi energy. The energy rate transfer is the product of the energy per collision, the scattering rate and the number of electrons

$$\frac{dE_e}{dt} = \frac{k_B^2 \theta_D D(E_F) v_F T_e (T_e - T_l)}{r_s T_0}. \tag{3.8}$$

where $T_l$ is replaced by $T_e - T_l$ to stop the energy transfer when the electronic and the atomic systems present the same temperature. This expression can be inserted into Eq. (3.6), the electron-phonon coupling being $G = \frac{k_B^2 \theta_D D(E_F) v_F T_e}{r_s T_0}$ and the electronic heat capacity $C_e = \frac{\pi^2}{3} k_B^2 D(E_F) T_e$. Another similar expression can be derived for the phonons. By ignoring the relatively slow phononic diffusion and considering $T_e$ constant, the ionic dynamics can be approximated by

$$\frac{\partial T_l}{\partial t} = -\alpha (T_e - T_l) \tag{3.9}$$

where $\alpha = \frac{3\theta_D C_e v_F}{\pi^2 r_s T_0 C_l}$ controls the cooling rate of the lattice due to the electrons When multiplied by $3k_B$, Eq. (3.9) gives the rate of energy transfer from a single atom to the electrons. In the simulations in [6], the authors confirm that Ni (that has a higher electron-phonon couling) quenches at a faster rate than Cu, as displayed in Fig. 3.1, highlighting that the electron-phonon coupling plays a crucial role in the cooling rate of thermal spikes.

The 2TM was applied to interpret the dynamics of a gold film that was heated by a laser in [7] (see also Sect. 2.2). For different fluences, no single value of $G$ could fit the data, while a single value of $G$ could be picked for long enough times when the subsystems were close to equilibration. The authors comment that the validity of the 2TM is restricted to a regime of a much faster electron-electron equilibration than the electron-phonon one. In their example, the electrons require a finite time to equilibrate and during this time there is a possibility of an increased heat transfer to the phonons due to non-thermal electrons.

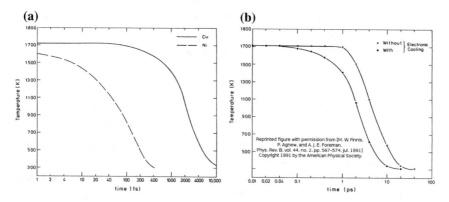

**Fig. 3.1** Temperature at the center of a 2 keV radiation cascade [6]. **a** shows that Ni quenches faster than Cu because of its stronger electron-phonon interaction. **b** shows the different cooling rates of Cu with and without the electronic contribution to cooling

Groeneveld et al. and Lisowski et al. [12, 13] proposed an improvement of the 2TM where the electron subsystem is split into a thermalized electron part and a non-thermalized one. The relative density of the former is described by $(1 - r)$, while the latter one is $r$. Thermalized electrons must be distributed according to a FD distribution, but there is no clear choice about the distribution of non-thermal ones. The authors discovered that another FD distribution with an auxiliary temperature $T_{nt}$ can effectively describe these electrons, so that the full electron population is described by

$$N(E) = ( (1 - r)f(T_e, E) + rf(T_{nt}, E) )\text{DOS}(E). \qquad (3.10)$$

The laser energy is directly injected into the non-thermal electrons and, as shown in Fig. 2.7a, at the experimental maximum laser fluence, the non-thermal part of the spectrum peaks at 18% of the total electron population. A comparison of this extended model with the standard 2TM and a time dependent FD fit of the experimental electron population is displayed in Fig. 3.2a. The 2TM predicts a maximum $T_e = 1300$ K and an equilibrium $T_e = 250$ K, while the experiment gives a peak of $T_e = 225$ K and an equilibrium $T_e = 135$ K. The extended model reproduces the temperature dynamics fairly well compared to the experiment. The authors concluded that the dynamics of energy flow must differ from the one in the 2TM. They believe that non-thermal electrons give rise to ballistic transport into the bulk, which cannot be described by diffusive models like the 2TM. The high energy non-thermal carriers are very efficient in ballistic and diffusive transport out of the excited spatial region [13].

The coupling of excited electrons to phonons is different from the coupling of thermalized ones [15]. Therefore a temperature dependent coupling strength can lead to a better agreement with experimental data. In [14] the authors apply the 2TM to warm dense copper created by the absorption of an ultrafast optical pulse. They use both a constant $G$ and a temperature dependent one from [16] and compare them with experimental data in Fig. 3.2b. Both temperatures from the 2TM peak at the end of the laser pulse and decrease in time, but the case with $G(T_e)$ cools down faster and agrees better with the experiment. This implies that in that specific temperature regime, the electron-phonon coupling is enhanced.

## 3.3 The 2TM in MD Simulations

To simulate atomically a projectile hitting matter, molecular dynamics simulations are usually employed. Adiabatic molecular dynamics simulations include atom-atom collisions that are very different compared to electron-atom ones where the energy and momentum exchange is much smaller than the typical atomic energy and momentum. Simple MD simulations lack a mechanism of atomic energy loss due to inelastic scattering with electrons. A possible way to include this electronic contribution is applying a continuous damping force to each ion $j$ [6]

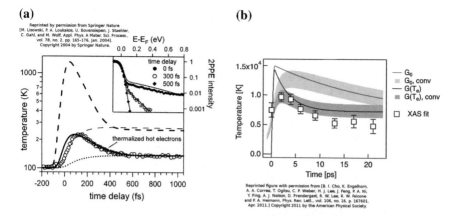

**(a)** **(b)**

**Fig. 3.2** In **a**, the dynamics of temperatures for optically excited Ru(001) [13]. In the standard 2TM, the electron temperature is a thick dashed line and the lattice temperature a thin dashed line, in the extended model the thermal electron temperature is a thick line and the lattice temperature a dotted line. The circles correspond to temperatures determined from FD fits of the thermal part of the electron spectra. In the inset, there are electron spectra at different delays and the relative spectra from the extended 2TM as lines. In **b**, the time evolution of the electron temperature [14] in excited warm dense copper. The upper line comes from the 2TM with a constant $G$, while for the lower line $G$ depends on the electronic temperature. The squares are from the experimental data and the shaded regions include the experimental accuracy into the 2TM results

$$F_j = -\beta_j v_j. \tag{3.11}$$

so that its Newtonian equation of motion becomes

$$m_j \frac{d^2 R_j}{dt^2} = F_j - \beta_j \frac{d R_j}{dt}. \tag{3.12}$$

The choice of the drag coefficient $\beta_J$ is essential to characterize the effect of the electrons on the atomic dynamics.

The projection of the force onto $v_j$ is the rate of energy loss and results from the interplay between friction and noise. By using Eq. (3.9), we have

$$-\beta_j |v_j|^2 = -3k_B \alpha (T_e - T_l). \tag{3.13}$$

Setting the thermal energy of ion $j$ equal to its kinetic energy $3k_B T_l = m_j v_j^2$, the drag coefficient becomes

$$\beta_j = \frac{3\theta_D C_e v_F m_j}{\pi^2 r_s T_0 C_l} \left( \frac{T_l - T_e}{T_l} \right). \tag{3.14}$$

Other forms of $\beta_j$ have been derived in [17, 18].

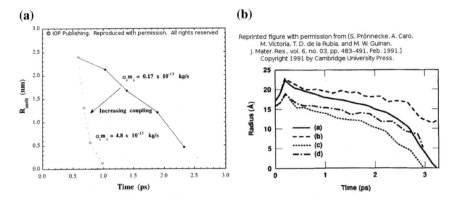

**(a)**

$\alpha_i m_i = 0.17 \times 10^{-13}$ kg/s

Increasing coupling

$\alpha_i m_i = 4.8 \times 10^{-13}$ kg/s

$R_{melt}$ (nm)

Time (ps)

**(b)**

Radius (Å)

— (a)
--- (b)
······ (c)
—·— (d)

Time (ps)

**Fig. 3.3** In **A**, the radius dynamics of the target molten region for two different couplings [22]. In **B**, the radius of the molten region of Cu as a function of time [23]. a and b represent a 5 keV cascade, with **a** a simulation with electronic coupling and **b** without. c and d a 2.5 keV cascade with c including the coupling and **d** excluding it

In [19–21], the authors used MD simulations with a constant $\beta$ to assess the effect of radiation cascades on semiconductors and metals and the influence of the electronic stopping power on it. In [22], the authors run a 2TM molecular dynamics simulation to estimate the cooling rate and the defect production for different couplings. They find that the cooling rate for strong couplings is much faster, as can be seen in Fig. 3.3a. For two identical radiation spikes, the case of strong coupling shows a complete disappearance of the molten region after 1 ps, whereas with low coupling the molten area is still present for times longer than 2 ps.

### 3.3.1  Augmented MD Models, the Langevin Equation

Compared to elastic atom-atom collisions, the relative importance of inelastic electron-atom scattering strongly depends on the energy of the projectile. In a study of Cu with Lindhard theory [24], it was estimated that, for projectile energies $\approx 10^6$ eV, the most important projectile energy loss mechanism was the electronic one (electronic stopping power (ESP) regime). At energies $\approx 10^4$ eV the electronic contribution decreased to 20–30%, while at $\approx 25$ eV it was only 8% of the total energy loss. The latter case corresponds to an electron-phonon interaction (EPI) regime.

Ideally, the electron-phonon coupling term in an MD simulation would be able to capture both regimes, but the wide range of energies from a few eV to MeV makes it a daunting challenge. One of the first ideas aimed at describing both regimes at the same time was the introduction of Langevin dynamics to model the electrons as a Langevin heat bath [25]. Electrons supply the atoms with both a kinetic loss mechanism and a random kick.

The atoms obey the equation of motion

$$m_j \frac{d^2 \mathbf{R}_j}{dt^2} = \mathbf{F}_j + \boldsymbol{\eta}_j(t) - \beta_j \frac{d \mathbf{R}_j}{dt} \tag{3.15}$$

where $\beta_j$ measures the coupling to the electron bath and $\boldsymbol{\eta}_j(t)$ is a stochastic force with the following average and time correlation properties[2]

$$\langle \boldsymbol{\eta}(t) \rangle = 0 \tag{3.16}$$

$$\langle \boldsymbol{\eta}(t) \cdot \boldsymbol{\eta}(t') \rangle = 2\beta k_B T_e \delta(t - t') \tag{3.17}$$

and this Gaussian probability distribution

$$P(\eta) = \left(2\pi \left\langle \eta^2 \right\rangle\right)^{-1/2} e^{-\frac{\eta^2}{2\langle \eta^2 \rangle}}. \tag{3.18}$$

To use this equation in a full radiation event, one must assume that the physics in both the ESP and the EPI regime is the same: the distinction between regimes would be determined by the different electron density explored by the projectile, as, in this picture, $\beta_j$ is a function of the local electronic density.

It is intuitive that an atom speeding across the lattice feels more drag contribution from the surrounding electronic clouds than a slow projectile that, after having lost most of its energy, bounces back and forth around a potential minimum. In [25], the authors picked a $\beta$ following linear response theory at high electronic densities and matching density functional theory at low densities.

An application of the Caro-Victoria model [25] is presented in [23], where the authors compare radiation heat spikes for different incoming energies in copper. In Fig. 3.3b they observe an increased cooling rate due to the presence of electronic coupling. As heat is transferred from the projectile to the electrons, $T_e$ should increase, changing in turn the random force $\eta$. In the Caro-Victoria model, the electronic heating is neglected as $T_e$ is a constant. To justify this with a rationale akin to the 2TM, for weak electron-phonon couplings, the electrons mean free path can be hundreds of Å. Because of their large thermal conductivity in metals, electrons can propagate heat very fast and in a large area of the target. Therefore, electronic heating can be neglected as a first approximation, because electrons act as a perfect heat sink.

Inhomogeneous models allow a transfer of heat from the electrons back into the lattice by introducing a space and time dependent $T_e$.

---

[2]From now on, we are going to drop the atomic index $j$ for notation clarity in this section. It will be reintroduced when needed.

### 3.3.2 Inhomogeneous Models

In [26], the authors Duffy and Rutherford deal with the energy feedback problem by coupling an MD simulation to an inhomogeneous electronic heat diffusion model. The main difference with the Caro-Victoria model is the use of an inhomogeneous Langevin thermostat that depends on space and time: the electronic temperature $T_e$ is local and is derived by integrating at each time step a thermal diffusion equation. Locality allows an energy feedback from the hot electrons back into the lattice, whereas homogeneous electron distributions are featureless and can only act as an energy sink where the heat from a thermal spike can dissipate.

The evolution of the atoms obeys an equation such as (3.15), where the stochastic force $\eta$ is defined as Eq. (3.17) and it depends on space through the locality of $T_e$. For choosing the damping coefficient $\beta$, the authors considered two limiting energy loss mechanisms. At high projectile energies, dissipation is dominated by a global stopping term $\beta_s$ that makes the energy loss proportional to the kinetic energy of the projectile. At low velocities, the electron-phonon loss $\beta_p$ is proportional to the temperature difference between the ionic and electronic system. Thus, with the introduction of a velocity cutoff $v_c$ to distinguish between regimes, the total drag coefficient on atom $j$ is

$$\begin{aligned} \beta_j &= \beta_p + \beta_s \quad v_j \geq v_c \\ &= \beta_p \qquad\quad v_j < v_c. \end{aligned} \tag{3.19}$$

The temporal and spatial evolution of $T_e$ is governed by a diffusion equation

$$c_e \frac{\partial T_e}{\partial t} = \nabla(k_e \nabla T_e) - g_p(T_e - T_l) + g_s T_l' \tag{3.20}$$

whose second term is the standard source from the 2TM (3.6). The authors partition space in several cells comprising a few hundred atoms $N$ and solve the heat diffusion equation using a finite difference method. In each cell $J$, the electronic temperature $T_e$ is a constant.

The lattice temperature $T_l$ is defined by averaging the kinetic energies of all atoms in a cell

$$\frac{3}{2} k_B T_l = \frac{1}{N} \sum_{j \in J} \frac{1}{2} m v_j^2, \tag{3.21}$$

while $T_l'$ is averaged only over the atoms with a $v_j \geq v_c$ that belong to $J'$

$$\frac{3}{2} k_B T_l' = \frac{1}{N} \sum_{j \in J'} \frac{1}{2} m v_j^2 \tag{3.22}$$

The couplings $g_p$ and $g_s$ are related to the electron-phonon coupling and the stopping power respectively. They are determined by energy balance equations

$$g_p = \frac{3Nk_B\beta_p}{\Delta V m} \qquad (3.23)$$

$$g_s = \frac{3N'k_B\beta_s}{\Delta V m} \qquad (3.24)$$

where $\Delta V$ is the volume of the cell and $N'$ the number of atoms with a velocity higher than the cutoff.

All the parameters $c_e$, $k_e$, $\beta_p$ and $\beta_s$ depend clearly on the material under investigation and could strongly depend on the lattice and electronic temperature, but in [26] they are chosen to be constant. The authors study the time evolution of the electronic and the atomic temperature for a low energy 10 keV cascade in Fe. In Fig. 3.4, they show the results testing a range of stopping and electron-phonon relaxation times $\tau_{s/p} = m/\beta_{s/p}$. Decreasing relaxation times correspond to increasing couplings. While the electronic temperature vary with the stopping parameter and increases with high $\beta_s$, the atomic temperature does not change significantly. For varying $\beta_p$, the lattice shows different cooling rates. Their preliminary analysis about defect production reveals that the number of defects decreases with higher

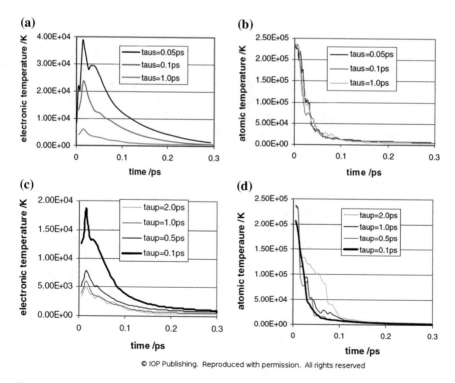

**Fig. 3.4** Simulation of a 10 keV cascade in Fe [26]. In **a** and **b**, time evolution of $T_e$ and $T_l$ for different $\tau_s$ and $\tau_p = 1$ ps; in **c** and **d**, a scan in $\tau_p$ and $\tau_s = 1$ ps

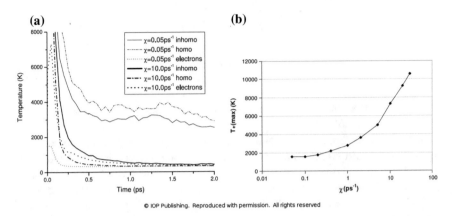

**Fig. 3.5** 10 keV radiation cascades on Fe from [27]. In **a**, the time evolution of the maximum temperature for two different electron-phonon couplings $\chi$ (a strong one at 10 ps$^{-1}$ and a weak one at 0.05 ps$^{-1}$) and different thermostats. The *homo* and *inhomo* plots refer to the maximum $T_l$ for respectively an homogeneous thermostat (Caro-Victoria) and an inhomogeneous one. The *electrons* plot is related to the maximum $T_e$ in the inhomogeneous case. In **b**, the maximum $T_e$ for different $\chi$

electron-phonon couplings, confirming Flynn's model [2] where high couplings corresponded to fast electronic heat transport away from the thermal spike region. Even for low energy radiation cascades, the electrons do influence the dynamics of heat transport, especially when the electron-phonon interaction is large.

In a following paper [27], Rutherford and Duffy perform a comparison of their inhomogeneous method with the homogeneous Caro-Victoria one. For representing the strength of the couplings, they use the inverse of the electron-phonon relaxation time $\chi = 1/\tau_p$. Figure 3.5a shows that the cooling rate changes significantly for different $\chi$: the system cools down much faster for strong couplings. The evolution of the electronic temperature can be split in three phases: a first one lasting a few tens of fs where the electrons heat up, one that lasts hundreds of fs when there is a fast cooling down due to electronic diffusion and a last one with a slow temperature decay related to exchanges of energy between electrons and ions.

Comparing the two thermostats, for weak couplings the atoms would move relatively fast, so the electronic stopping introduced in the inhomogeneous case would be effective. In fact, the atomic temperature for weak couplings is lower in the inhomogeneous case. On the other hand, for strong couplings the inhomogeneous simulation presents a higher $T_l$ than the homogeneous case. In the former, the presence of a high electronic temperature and its feedback into the lattice compensates for the increased loss of energy due to the stopping power and causes a lower cooling rate. In Fig. 3.5b it is shown that the maximum electronic temperature increases significantly for strong couplings. So a high $T_e$ can be found even in low energy cascades if the coupling is intense enough.

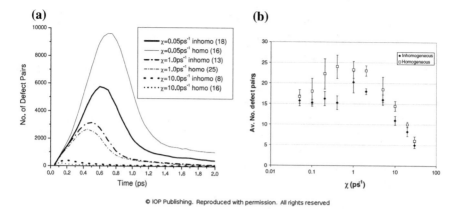

**Fig. 3.6** In **a**, time evolution of the number of defects for three increasingly strong couplings 0.05 ps$^{-1}$, 1 ps$^{-1}$, 10 ps$^{-1}$ and both the homogeneous and inhomogeneous thermostat. The number in brackets in the legend is the number of stable defects at the end of the simulation. In **b**, the average number of stable defects pair caused by different couplings for the homogeneous and inhomogeneous thermostat [27]

Regarding the formation of defects, it is clear from Fig. 3.6a that a strong coupling leads in general to a smaller number of defects compared to a weak coupling. This is due to the fast energy dissipation mechanism provided by the electrons which cause a quick quenching of the thermal spike. A comparison of the thermostats shows that for all couplings the peak number of defects is lower for the inhomogeneous thermostat because of the additional dissipation of the stopping electronic term. The high electronic temperature plays an important role contributing to disorder and enlarging the molten region.

The residual defects number, from Fig. 3.6b, is lower for the inhomogeneous thermostat compared to the homogeneous case. The slowed down cooling of the thermal spike of the inhomogeneous simulations is caused by the energy feedback from the hot electrons which has an enhanced annealing effect. This fact suggests that materials with strong electron-ion coupling could display an enhanced radiation damage resistance. A recent article [28] confirms these findings for cubic silicon carbide.

A recent publication [29] studies a radiation cascade in Fe with very high PKA energies (0.2–0.5 MeV) for a very large system of 100–500 million atoms. The authors discover that most of the damage made in the first ps is repaired through diffusion and recombination processes in a few tens of ps. They do not observe any branching of the collision cascade with the formation of areas of damage well separated from each other except for the very first moments of the collision. Subcascades tend to recover well in materials resistant to amorphization like metals. The authors give a qualitative explanation for the absence of branching and the presence of a continuous damage region: a significant displacement of an atom from its equilibrium position requires a lot of energy and, neglecting inelastic energy losses, the transferred energy in a collision is $T \approx 1/E^2$ where $E$ is the incoming atom energy. Large $E$ correspond to small $T$ and a more localized damage region. They note that for a decreasing $E$, $T$ increases and sub cascade branches develop.

Even though inhomogeneous models are generally more realistic than adiabatic MD models because they include energy feedback from electrons to atoms, they still treat the electronic subsystem in an implicit semi-classical way. Electrons appear as a medium where the atoms are immersed. The distinction between electron stopping and electron-phonon regime is a semi-empirical approach to describe what is effectively single phenomenon. A way to improve on this would require a significant leap both from the theoretical and from the computational point of view: treating the electrons explicitly and individually.

# References

1. Ziman, J.M. 1972. *Transport properties, in principles of the theory of solids*, 211–254. Cambridge: Cambridge University Press. https://doi.org/10.1017/CBO9781139644075.009.
2. Flynn, C.P., and R.S. Averback. 1988. Electron-phonon interactions in energetic displacement cascades. *Physical Review B* 38 (10): 7118. https://doi.org/10.1103/PhysRevB.38.7118.
3. de la Rubia, T.D., R.S. Averback, R. Benedek, and W.E. King. 1987. Role of thermal spikes in energetic displacement cascades. *Physical Review Letters* 59 (17): 1930–1933. https://doi.org/10.1103/PhysRevLett.59.1930.
4. Kaganov, M.I., I.M. Lifshitz, and L.V. Tanatarov. 1957. Relaxation between electrons and the crystalline lattice. *Soviet Physics JETP* 4: 173–178.
5. Anisimov, S.I., B.L. Kapeliovich, and T.L. Perel'man. 1975. Electron emission from metal surfaces exposed to ultrashort laser pulses. *Journal of Experimental and Theoretical Physics* 39: 375–377.
6. Finnis, M.W., P. Agnew, and A.J.E. Foreman. 1991. Thermal excitation of electrons in energetic displacement cascades. *Physical Review B* 44 (2): 567–574. https://doi.org/10.1103/PhysRevB.44.567.
7. Fann, W.S., R. Storz, H.W.K. Tom, and J. Bokor. 1992. Direct measurement of nonequilibrium electron-energy distributions in subpicosecond laser-heated gold films. *Physical Review Letters* 68 (18): 2834–2837. https://doi.org/10.1103/PhysRevLett.68.2834.
8. Toulemonde, M., C. Dufour, and E. Paumier. 1992. Transient thermal process after a high-energy heavy-ion irradiation of amorphous metals and semiconductors. *Physical Review B*, 46 (22): 14 362–14 369. https://doi.org/10.1103/PhysRevB.46.14362.
9. Toulemonde, M., W.J. Weber, G. Li, V. Shutthanandan, P. Kluth, T. Yang, Y. Wang, and Y. Zhang. 2011. Synergy of nuclear and electronic energy losses in ion-irradiation processes: The case of vitreous silicon dioxide. *Physical Review B* 83 (5): 054106. https://doi.org/10.1103/PhysRevB.83.054106.
10. Tzou, D. 2014. *Heat transport by phonons and electrons, in macro- to microscale heat transfer*, 1–59. Chichester, UK: John Wiley & Sons Ltd. https://doi.org/10.1002/9781118818275.ch1.
11. Race, C. 2011. The treatment of electronic excitations in atomistic simulations of radiation damage–a brief review. In *The modelling of radiation damage in metals using ehrenfest dynamics*. Springer Theses. Springer, https://doi.org/10.1007/978-3-642-15439-3_3.
12. Groeneveld, R.H.M., R. Sprik, and A. Lagendijk. 1995. Femtosecond spectroscopy of electron-electron and electron-phonon energy relaxation in Ag and Au. *Physical Review B* 51 (17): 11 433–11 445. https://doi.org/10.1103/PhysRevB.51.11433.
13. Lisowski, M., P. Loukakos, U. Bovensiepen, J. Stähler, C. Gahl, and M. Wolf. 2004. Ultrafast dynamics of electron thermalization, cooling and transport effects in Ru(001). *Applied*

*Physics A: Materials Science & Processing* 78 (2): 165–176. https://doi.org/10.1007/s00339-003-2301-7.

14. Cho, B.I., K. Engelhorn, A.A. Correa, T. Ogitsu, C.P. Weber, H.J. Lee, J. Feng, P.A. Ni, Y. Ping, A.J. Nelson, D. Prendergast, R.W. Lee, R.W. Falcone, and P.A. Heimann. 2011. Electronic structure of warm dense copper studied by ultrafast X-ray absorption spectroscopy. *Physical Review Letters* 106 (16): 167601. https://doi.org/10.1103/PhysRevLett.106.167601.

15. Mueller, B.Y., and B. Rethfeld. 2013. Relaxation dynamics in laser-excited metals under nonequilibrium conditions. *Physical Review B* 87 (3): 035139. https://doi.org/10.1103/PhysRevB.87.035139.

16. Lin, Z., L.V. Zhigilei, and V. Celli. 2008. Electron-phonon coupling and electron heat capacity of metals under conditions of strong electron-phonon nonequilibrium. *Physical Review B* 77 (7): 075133. https://doi.org/10.1103/PhysRevB.77.075133.

17. Montgomery, M.J., T.N. Todorov, and A.P. Sutton. 2002. Power dissipation in nanoscale conductors. *Journal of Physics: Condensed Matter* 14 (21): 5377–5389. https://doi.org/10.1088/0953-8984/14/21/312.

18. Lü, J.-T.T., M. Brandbyge, P. Hedegård, T.N. Todorov, D. Dundas, P. Hedegard, T.N. Todorov, and D. Dundas. 2012. Current-induced atomic dynamics, instabilities, and Raman signals: Quasiclassical Langevin equation approach. *Physical Review B* 85 (24): 245444. https://doi.org/10.1103/PhysRevB.85.245444.

19. Nordlund, K., M. Ghaly, and R.S. Averback. 1998. Mechanisms of ion beam mixing in metals and semiconductors. *Journal of Applied Physics* 83 (3): 1238. https://doi.org/10.1063/1.366821.

20. Nordlund, K., L. Wei, Y. Zhong, and R. Averback. 1998. Role of electron-phonon coupling on collision cascade development in Ni, Pd, and Pt. *Physical Review B* 57 (22): R13 965–R13 968. https://doi.org/10.1103/PhysRevB.57.R13965.

21. Nordlund, K., M. Ghaly, R. Averback, M. Caturla, T. Diaz de la Rubia, and J. Tarus. 1998. Defect production in collision cascades in elemental semiconductors and fcc metals. *Physical Review B* 57 (13): 7556–7570. https://doi.org/10.1103/PhysRevB.57.7556.

22. Gao, F., D.J. Bacon, P.E.J. Flewitt, and T.A. Lewis. 1998. The effects of electron-phonon coupling on defect production by displacement cascades in -iron. *Modelling and Simulation in Materials Science and Engineering* 6 (5): 543–556. https://doi.org/10.1088/0965-0393/6/5/003.

23. Prönnecke, S., A. Caro, M. Victoria, T.D. de la Rubia, and M. Guinan. 1991. The effect of electronic energy loss on the dynamics of thermal spikes in Cu. *Journal of Materials Research* 6 (03): 483–491. https://doi.org/10.1557/JMR.1991.0483.

24. Lindhard, J., and M. Scharff. 1961. Energy dissipation by ions in the kev region. *Physical Review* 124 (1): 128–130. https://doi.org/10.1103/PhysRev.124.128.

25. Caro, A., M. Victoria. 1989. Ion-electron interaction in molecular-dynamics cascades. *Physical Review A* 40 (5): 2287–2291. https://doi.org/10.1103/PhysRevA.40.2287.

26. Duffy, D.M., and A.M. Rutherford. 2007. Including the effects of electronic stopping and electron-ion interactions in radiation damage simulations. *Journal of Physics: Condensed Matter* 19 (1): 016207. https://doi.org/10.1088/0953-8984/19/1/016207.

27. Rutherford, A.M., and D.M. Duffy. 2007. The effect of electron-ion interactions on radiation damage simulations. *Journal of Physics: Condensed Matter* 19 (49): 496201. https://doi.org/10.1088/0953-8984/19/49/496201.

28. Zhang, C., F. Mao, and F.-S. Zhang. 2013. Electron-ion coupling effects on radiation damage in cubic silicon carbide. *Journal of physics. Condensed matter: an Institute of Physics journal* 25 (23): 235402. https://doi.org/10.1088/0953-8984/25/23/235402.

29. Zarkadoula, E., S.L. Daraszewicz, D.M. Duffy, M.A. Seaton, I.T. Todorov, K. Nordlund, M.T. Dove, and K. Trachenko. 2013. The nature of high-energy radiation damage in iron. *Journal of physics. Condensed matter: an Institute of Physics journal* 25 (12): 125402. https://doi.org/10.1088/0953-8984/25/12/125402.

# Chapter 4
# Simulating Electrons and Phonons: Atomistic Methods

In this chapter we discuss explicit methods that tackle electron-phonon problems at the microscopic level. To capture the *exact* dynamics of coupled systems of interacting electrons and phonon modes is an insurmountable task for computer simulations. The tremendous memory footprint of the full problem makes simulations unfeasible even for systems including just a few electrons. Approximations must be employed and some information has to be discarded. Depending on the problem under study, it is common to focus only on certain aspects of the full time-dependent coupled problem.

Time independent methods can be employed to study static quantities, as electron-phonon scattering rates. One of the simplest approaches is based on the Fermi Golden Rule, where the electron-phonon interaction is included as a first order perturbative correction. This qualitative approach can provide preliminary insights, but it cannot describe strong couplings where multiple scattering events are relevant. A standard generalization that allows the description of higher order processes is based on Non-Equilibrium Green's Functions (NEGF) [1], in which the Dyson equation is expanded in a Born series and where some low energy Feynman diagrams are included in the self-energy expression. However the applicability of NEGF is limited by the large amount of computer power required to simulate even very small systems. Another method based on Green's functions is the Self-Consistent Born Approximation which has been widely used for modelling inelastic transport [2, 3], also in time dependent problems [4].

The study of non-adiabatic phenomena with possible connections with MD requires time-dependent methods. Ehrenfest dynamics and its extension Correlated Electron-Ion Dynamics (CEID) are dynamical methods that discard a part of the electron-phonon correlation to make non-adiabatic problems less computationally demanding. They can estimate physical quantities as atomic trajectories that can

© Springer International Publishing AG, part of Springer Nature 2018
V. Rizzi, *Real-Time Quantum Dynamics of Electron-Phonon Systems*,
Springer Theses, https://doi.org/10.1007/978-3-319-96280-1_4

be connected to MD simulations. A reworking of the CEID method that converges systematically even for strong couplings is described in [5, 6].

In this chapter we focus on a selection of methods that are especially relevant for their connections with our work.

## 4.1   A Simple Classical Model and the Born-Oppenheimer Approximation

A simple classical model from [7] is a stimulating starting point for the discussion. Even though qualitative and unrealistic, it can offer important insights on the physical features of electron-phonon phenomena. In this model both the atoms and the electrons are treated classically and they evolve according to Newton's equation of motion. A single atom $A$ with mass $M$ is immersed in a lattice and interacts with its neighbours through harmonic oscillators with a spring constant $K$. The electrons do not interact with each other but are coupled to atom $A$. The full Hamiltonian of the system is

$$H = H_e + H_A + H_{eA} \tag{4.1}$$

$$= \underbrace{\sum_i \left( \frac{p_i^2}{2m} + v(r_i) \right)}_{H_e} + \underbrace{\left( \frac{P^2}{2M} + \frac{1}{2}KX^2 \right)}_{H_A} - \underbrace{\sum_i X \cdot \nabla_i v(r_i)}_{H_{eA}} \tag{4.2}$$

where $m$ is the electron's mass, $r_i$ and $p_i$ are the position the momentum of electron $i$, $P$ and $X$ are the momentum and the displacement of the moving atom. For small $X$, the electronic potential can be approximated by using $v(r_i - X) = v(r_i) - X \cdot \nabla_i v(r_i)$. By using Hamilton's equations[1] on Eq. (4.1), we obtain a set of EOM for the electrons and the atom.

In the trivial case where the coupling $H_{eA}$ is switched off, the EOM are decoupled and have this simple form

$$\ddot{X}^k = -\frac{K}{M}X^k \tag{4.3}$$

$$\ddot{r}_i^k = -\frac{1}{m}\frac{\partial v(r_i)}{\partial r_i^k}. \tag{4.4}$$

The resulting atomic motion is harmonic $X^k(t) = A^k \sin(\sqrt{K/M}t + \phi^k)$, while the dynamics of the electrons unfolds on a constant energy surface. The full EOM with $H_{eA}$ are

---

[1]For the atom they read $\dot{X}^k = \frac{\partial H}{\partial P^k}$, $\dot{P}^k = -\frac{\partial H}{\partial X^k}$ while for the $i$th electron $\dot{r}^k = \frac{\partial H}{\partial p^k}$, $\dot{p}^k = -\frac{\partial H}{\partial r^k}$. Index $k$ refers to the $x$, $y$, $z$ atomic components.

$$\ddot{X}^k = -\frac{K}{M}X^k + \frac{1}{M}\sum_i \frac{\partial v(r_i)}{\partial r_i^k} \tag{4.5}$$

$$\ddot{r}_i^k = -\frac{1}{m}\frac{\partial v(r_i)}{\partial r_i^k} + \frac{1}{m}\sum_j X^j \frac{\partial^2 v(r_i)}{\partial r_i^k \partial r_i^j} \tag{4.6}$$

and do not allow an explicit solution any more because the atomic motion depends on the dynamics of all the electrons. Moreover, the electrons are implicitly coupled to each other by $X$, that depends on *all* the electrons. The last term in Eq. (4.6) acts as an external time-dependent force on the electrons and makes their energy change.

In [8], the authors solved the EOM in 1D in the limit $m \ll M$ where the atom behaves as a hard wall for the light electrons. The power $w$ delivered to the atom by a current $j$ is

$$w \sim 4j\left(\frac{m}{M}\right)(K_e - 2K_A) \tag{4.7}$$

where $K_e$ and $K_A$ are the average electronic and atomic kinetic energies. The opposite signs of the terms allow both cooling and heating effects, depending on the magnitude of the kinetic energies. For very small biases, the electrons would be slow and cold, so $K_e$ would be very small and there would be a cooling of the atom. On the other hand, for high biases the electrons would cause a heating up of the atom.

The model allows to observe the microscopic correlated electron nuclear fluctuations caused by inelastic exchanges of energy between the electrons and the atom. Following the approximations in [7] and in the limit of a characteristic electronic frequency much larger than the atomic one, the average kinetic energy of the atom would be $K_A = \frac{3}{2}\frac{m}{M}K_e$. The heavy atom would be moving because of the correlated oscillations induced by the much lighter electron.

The significant mass difference between electrons and atoms determines a large timescale difference between them which has been exploited widely and is the basis for one of the most common approximations in solid state physics: the Born-Oppenheimer approximation (BOA). By the time an atom moves, the fast moving electrons have undergone sufficiently many collisions to minimize the system's free energy. Because of this, it is natural to assume that for any nuclear position, the electrons occupy instantaneously their ground state. This means that the atomic system is effectively decoupled from the electronic dynamics: the electrons rearrange depending on the instantaneous atomic position, while the atoms move in a potential determined by the electronic ground state. This approximation greatly simplifies the complexity of the problem that lies in the electron-atom coupled motion.

The BOA is often used to determine transition rates $\Gamma$ in conjunction with the Fermi Golden Rule

$$\Gamma_{i\rightarrow f} = \frac{2\pi}{\hbar}\left|\langle f|\hat{H}_I|i\rangle\right|^2 \delta(E_f - E_i) \tag{4.8}$$

where the final state energy is $E_f = \langle f|\hat{H}_0|f\rangle$, the initial state energy is $E_i = \langle i|\hat{H}_0|i\rangle$, $\hat{H}_0$ is the unperturbed Hamiltonian and $\hat{H}_I$ is the perturbation that causes the transition from an initial state $|i\rangle$ to final state $|f\rangle$,.

For example, in [7], the system's Hamiltonian is treated perturbatively around small atomic displacements to derive atomic power

$$w = \frac{2\pi}{\hbar} \sum_{n'N'} \left|\langle\Psi_{n'N'}|\hat{H}_{eN}|\Psi_{nN}\rangle\right|^2 (W_{N'} - W_N)\delta(U_{n'N'} - U_{nN}) \qquad (4.9)$$

where $W_N$ represents the energy of the nuclei in state $N$, state $\Psi_{nN} = |i\rangle$, $\Psi_{n'N'} = |f\rangle$, $U_{nN} = E_i$ and $U_{n'N'} = E_f$. This perturbative model has been applied to a vast number of different systems. In [8], the authors prove that in the high bias limit, the perturbative approach gives results in approximate agreement with those of the classical model above.

## 4.2 Ehrenfest Dynamics

An intrinsic drawback of the BOA is its inability to describe non-adiabatic phenomena where the mutual interaction between electrons and atoms plays a crucial role on each other's dynamics. For example, during the first phase of a radiation cascade, for sufficiently energetic incoming particles, the electrons along the track would be excited out of their ground state. At successive times, both the electronic and the atomic configuration would be changing, violating the adiabatic assumption of the BOA. A common approach that includes non-adiabatic features is Ehrenfest dynamics (ED) [9, 10]. Its main assumptions are a mean field approach to the electron-nuclear interaction and a classical treatment of the atoms.

ED starts from Hamiltonian $\hat{H}(\hat{R}) = \hat{K}_A + \hat{H}_e(\hat{R})$ where the kinetic contribution of the nuclei $\hat{K}_A$ has been separated from $\hat{H}_e(\hat{R})$ that contains the kinetic energy of the electrons, the electron-electron, electron-nucleus and nucleus-nucleus interaction. Taking the expectation value of the nuclear position and momentum operators of the $\nu$-th nucleus[2] over full system's wavefunction $\bar{R}_\nu = \langle\Psi|\hat{R}_\nu|\Psi\rangle$, $\bar{P}_\nu = \langle\Psi|\hat{P}_\nu|\Psi\rangle$ and evolving them with the time-dependent Schrödinger equation, we get a set of *exact* equations, the Ehrenfest equations

$$\frac{d\bar{R}_\nu}{dt} = \frac{\bar{P}_\nu}{M_\nu} \qquad (4.10)$$

$$\frac{d\bar{P}_\nu}{dt} = \bar{F}_\nu \qquad (4.11)$$

---

[2]These operators satisfy the usual commutation rule $[\hat{R}_\nu, \hat{P}_{\nu'}] = i\hbar\delta_{\nu\nu'}$. Quantities such as $\hat{R}$ or $\bar{R}$ mean the collection of $\{\hat{R}_\nu\}$ and $\{\bar{R}_\nu\}$ for every $\nu$, that runs over the nuclei and the coordinate set.

where $M_\nu$ is the nuclear mass and the force is $\bar{F}_\nu = \langle \Psi | -\frac{\partial \hat{H}(\hat{R})}{\partial \hat{R}_\nu} | \Psi \rangle$. These equations look classical, but enclose the quantum complexity in the expectation value over the full system's wavefunction $|\Psi\rangle$.

ED can be generalized to a density matrix (DM) formalism. The expectation values above are replaced by traces, as $\bar{R}_\nu = Tr(\hat{\rho}\hat{R}_\nu)$, where $\hat{\rho}$ is the DM of the whole system and the electronic DM is $\hat{\rho}_e = Tr_A(\hat{\rho})$. The system's dynamics is then described by the quantum Liouville equation

$$i\hbar \frac{d\hat{\rho}}{dt} = [\hat{H}, \hat{\rho}]. \tag{4.12}$$

The first assumption of the Ehrenfest approximation is decoupling the system into a product of its nuclear and electronic components $\hat{\rho} = \hat{\rho}_A \otimes \hat{\rho}_e$. This decoupling causes a loss of correlation between the electronic and nuclear system. The second assumption is a classical treatment of the nuclei, which occupy classical point-like positions without any quantum width. Therefore, the nuclear part of the density matrix is centred over a single classical trajectory $\langle R|\hat{\rho}|R \rangle = \hat{\rho}_e \delta(R - \bar{R})$. With these approximations, the evolution of the electronic DM follows a Liouville equation

$$i\hbar \frac{d\hat{\rho}_e}{dt} = [\hat{H}_e(\bar{R}), \hat{\rho}_e] \tag{4.13}$$

where $\hat{H}_e(\bar{R}) = Tr_A\left(\hat{H}_e(\hat{R})\,\delta(\hat{R} - \bar{R})\right)$ is a mean field Hamiltonian that crucially depends on the expectation value of the nuclear positions $\bar{R}$. The evolution of the classical nuclear coordinates is still determined by Eqs. (4.10 and 4.11), while the expression for $\bar{F}_\nu$ simplifies into $\bar{F}_\nu \simeq -Tr_e\left(\hat{\rho}_e \frac{\partial \hat{H}_e(\bar{R})}{\partial \bar{R}_\nu}\right)$. Here, electrons respond to *individual* classical nuclear potentials, while nuclei interact with an average density of electrons, as in the BOA.

In [11], the authors test the effectiveness of classical models with a damping coefficient independent of the velocity of the nucleus by simulating radiation cascades with both implicit classical methods and a semi-classical time dependent tight binding (TDTB) method based on Ehrenfest approximation. They compare non-adiabatic energy transfers from atomic motion to electrons, as shown in Fig. 4.1. Their results show that a velocity *independent* damping coefficient is already a good first approximation for the nonadiabatic force, especially in high energy cascades. The model with a simple kinetic energy cutoff doesn't look to be in agreement with the Ehrenfest result, while the density dependent damping gives the closest results to the Ehrenfest calculation.

Ehrenfest dynamics can be combined with Time-dependent Density Functional Theory (TDDFT) and it has been used to study a projectile dynamics in the regime when it is losing energy to the electrons along its path [15]. This approach describes in a single simulation the projectile energy loss mechanism due to the electrons and the nuclear dynamics in the presence of the electrons excited by the projectile. Simulations are used to extract the non-adiabatic forces that act on the projectile

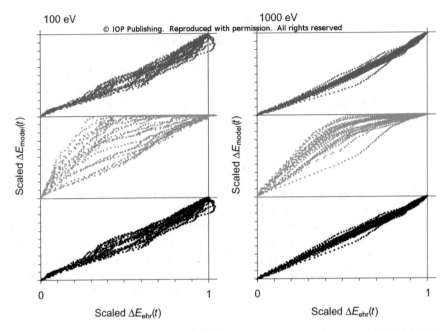

**Fig. 4.1** Plots of irreversible energy losses for a 100 eV cascade (left) and 1 keV (right) [11]. Three classic models on the y axis are compared with the Ehrenfest method on the x axis. A straight line would mean that classical and Ehrenfest losses coincide. The three classical methods are: a homogeneous damping as in [12] (top), a model with a kinetic cutoff like [13] (middle), an inhomogeneous electron density dependent damping as in [14] (bottom)

and depend in general on its velocity. In the case of a proton tunnelling through aluminium, the authors find that at high projectile velocities there is a strong radial momentum transfer to the surrounding nuclei, several times larger than the momentum transfer predicted by the adiabatic approximation. This non-adiabatic effect is caused by the electrons in the lattice that are not able to provide chemical bonding after being excited by the incoming radiation.

The same method has been used to determine the stopping power of hydrogen and helium through a gold lattice [16], comparing it with experimental results, as in Fig. 4.2. This work has helped to understand experimental features that could not be explained with established theories such as the nonlinear behaviour of the stopping power versus projectile velocity and the low projectile velocity regime. The deep lying states in the d-band of gold play a crucial role in determining the stopping power, even at low velocities, either by being excited directly or by providing extra screening. A recent study used Ehrenfest TDDFT to investigate the electronic stopping of hydrogen in germanium, a small band-gap semi-conductor [17].

If ED is effective in describing the electronic stopping phase of a radiation event, it cannot capture the electronic transport of energy away from the projectile and their subsequent thermalization with the nuclei via electron-phonon interaction [15]. Its

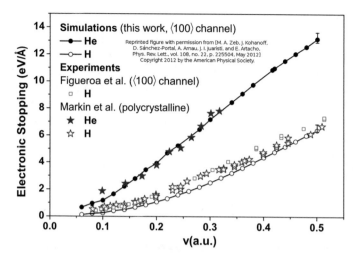

**Fig. 4.2** Electronic stopping power of hydrogen and helium projectiles versus projectile velocity. Ehrenfest TDDFT simulations [16] are compared to experiments

main limitation is the mean field description of the electrons that breaks the mutual heat exchange between electrons and nuclei, as we will discuss it in the next section.

### 4.2.1  A One-Sided Electron-Atom Heat Exchange

In [8], the authors use tight binding with ED to simulate a single atom moving in an infinite wire. They apply a bias to the wire's ends so that an electric current flows through the system and they find out that, for low biases, the oscillating atom cools down. This can be physically understood because, in this case, the injected electrons occupy low energy states in the wire and don't have enough energy to excite the atom. If we call $\omega$ the frequency of the lowest vibrational mode of the atom, the bias has to exceed $\hbar\omega$ to allow the electrons to start exchanging energy with the atom. The higher the bias, the more vibrational excitations get accessible with the possibility of activating multiple phonon processes. Therefore, for larger biases and electronic currents, the electrons are expected to heat up the atom, but ED is found to cool the atom down even at large biases.

This unphysical phenomenon is caused by an intrinsic limitation of the Ehrenfest approximation: it describes the electrons via an average charge density like a fluid, suppressing the fluctuations induced on the atoms by the moving electrons. The electronic heating caused by the atoms is captured correctly because the atoms appear explicitly in the electron dynamics. Atomic fluctuations can be related to an atomic temperature to which the electrons respond. The converse transfer of energy is missing because the atoms interact with a mean field electronic density which lacks microscopic details. The average electronic density cannot produce neither a mean force nor a force noise on the atomic motion. The electronic microscopic noise that is neglected by ED is the key for vibrational heating.

Atoms are not unmovable objects when the light electrons buzz around them: in a classical context, after each collision, an atom recoils and exchange a small amount of energy with the colliding electron, as the simple model from [7] highlighted in Sect. 4.1. Capturing these correlated electron-ion fluctuations is beyond the scope of the BOA and the Ehrenfest approximation.

Taking a classical electronic density $\rho_e$, with the electrons possessing a number of classical positions $x$ and momenta $p$, the atomic Ehrenfest force would be

$$\bar{F}_\nu = - \int dr\, dp\, \rho_e(r, p) \frac{\partial H_e(\bar{R}; r, p)}{\partial \bar{R}_\nu}. \tag{4.14}$$

Picking a single trajectory $\bar{r}$ for the electrons would make $\rho_e$ localized in phase space. Thus the atomic force would become $\bar{F}_\nu = -\frac{\partial \hat{H}_e(\bar{R}; \bar{r}, \bar{p})}{\partial \bar{R}_\nu}$. The integral in Eq. (4.14) can be seen as the average force produced by *all* the electronic trajectories, each with a weight $dr\, dp\, \rho_e(r, p)$. The averaging procedure hides away the microscopic correlations of the interaction and inhibits the flow of energy from the electrons to the atoms.

The simple model from [7] helps to understand the physical importance of these correlated fluctuations. Consider a system made of a single nucleus immersed in an electron cloud. Suppose that the nucleus experiences two kinds of forces: an harmonic one $-kX$ and a small non-adiabatic time-dependent one $f(t)$. The instantaneous power given to the nucleus is $\dot{U}(t) = \dot{X} \cdot f(t)$. Averaging this power over a period of time $\tau$, we get

$$\langle \dot{U} \rangle_\tau = \frac{1}{\tau} \int_{t_0}^{t_0+\tau} dt\, \dot{X} \cdot f(t). \tag{4.15}$$

For a pure harmonic motion, the time average of $\dot{X}$ would be null. For an $f(t)$ varying slowly in time, the motion of the atom would be mainly harmonic and the power would be very small. If $f(t)$ fluctuated quickly in time without being correlated with $\dot{X}$, the time integral in Eq. (4.15) would again average to zero. On the other hand, if $f(t)$ depended on $\dot{X}$ or if $\dot{X}$ was allowed to be perturbed by $f$, then the correlated fluctuations would not average to zero in the integral and only then they would produce a non-zero power. These correlations would disappear in the Ehrenfest approximation because of the average electronic trajectory acting on the nuclei.

In some problems, these correlations do not play an important role and can be safely ignored. In other problems, they are essential to capture physically important phenomena, for example Joule heating. A striking case of the ED failure is a molecule in an electronically excited state, with atoms relaxed on that Born-Oppenheimer surface. ED would produce no evolution at all, whereas in reality non radiative de-excitation can take place. The desire to be able to simulate a wide range of non adiabatic processes motivates the development of more general methods that can capture some electron-phonon correlation, while retaining the good qualities of the Ehrenfest method (i.e. the classical description of the nuclei and the possibility to treat the electrons via a single particle density).

## 4.3 Correlated Electron-Ion Dynamics (CEID)

In ED atoms are treated as classical particles, but they possess a quantum nature, therefore individual trajectories cannot fully describe atomic dynamics. Atoms can be modelled by wavepackets, as shown in Fig. 4.3, where the atom has an average position $\bar{R}_\nu$ and a packet spread $W$ that represents the uncertainty on its position. The quantum uncertainty of a particle is inversely proportional to its mass. Atomic masses are usually large, therefore $W$ is small compared to typical interatomic distances.

In light of this observation, the exact Ehrenfest equations

$$\frac{d\bar{R}_\nu}{dt} = \frac{\bar{P}_\nu}{M_\nu}, \qquad \frac{d\bar{P}_\nu}{dt} = \bar{F}_\nu, \qquad \bar{F}_\nu = -Tr\left(\hat{\rho}\frac{\partial \hat{H}_e}{\partial \hat{R}_\nu}\right) \qquad (4.16)$$

can be interpreted classically, by considering the atomic wavepacket to be localized on $R = \bar{R}_\nu$ and $P = \bar{P}_\nu$. Because of the small wavepacket width, it is reasonable to expand the atomic coordinates perturbatively, with the atomic position being $\hat{R}_\nu = \bar{R}_\nu + \Delta\hat{R}_\nu$ and the momentum $\hat{P}_\nu = \bar{P}_\nu + \Delta\hat{P}_\nu$. This assumption is the foundation of the Correlated Electron-Ion Dynamics (CEID) method [9, 10].

With the perturbative expansion in the atomic position, the electronic Hamiltonian becomes

**Fig. 4.3** Atomic wavepacket that shows the CEID approach to nuclear position [7]. In ED the postion of an atom is described by a delta function centred in $R = \bar{R}$, in CEID it is represented by a wavepacket with a finite width $W$

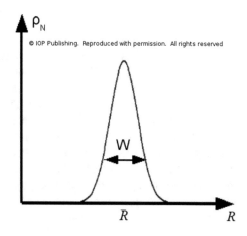

$$\hat{H}_e(\hat{R}) = \hat{H}_e(\bar{R}) + \sum_\nu \frac{\partial \hat{H}_e(\bar{R})}{\partial \bar{R}_\nu} \, \Delta \hat{R}_\nu + \frac{1}{2!} \sum_{\nu\nu'} \frac{\partial^2 \hat{H}_e(\bar{R})}{\partial \bar{R}_\nu \partial \bar{R}'_\nu} \, \Delta \hat{R}_\nu \Delta \hat{R}_{\nu'}$$

$$+ \frac{1}{3!} \sum_{\nu\nu'\nu''} \frac{\partial^3 \hat{H}_e(\bar{R})}{\partial \bar{R}_\nu \partial \bar{R}_{\nu'} \partial \bar{R}_{\nu''}} \, \Delta \hat{R}_\nu \Delta \hat{R}_{\nu'} \Delta \hat{R}_{\nu''} + \dots \tag{4.17}$$

$$= \hat{H}_e(\bar{R}) - \sum_\nu \hat{F}_\nu \, \Delta \hat{R}_\nu + \frac{1}{2} \sum_{\nu\nu'} \hat{K}_{2,\nu\nu'} \, \Delta \hat{R}_\nu \Delta \hat{R}_{\nu'} + \frac{1}{3!} \sum_{\nu\nu'\nu''} \hat{K}_{3,\nu\nu'\nu''} \, \Delta \hat{R}_\nu \Delta \hat{R}_{\nu'} \Delta \hat{R}_{\nu''} + \dots$$

$$\tag{4.18}$$

where the force operator is defined as $\hat{F}_\nu = -\frac{\partial \hat{H}_e(\bar{R})}{\partial \bar{R}_\nu}$, the spring operator as $\hat{K}_{2,\nu\nu'} = \frac{\partial^2 \hat{H}_e(\bar{R})}{\partial \bar{R}_\nu \partial \bar{R}'_\nu} = \frac{\partial \hat{F}_\nu}{\partial \bar{R}'_\nu}$ and the anharmonic spring operator as $\hat{K}_{3,\nu\nu'\nu''}$ $= \frac{\partial^3 \hat{H}_e(\bar{R})}{\partial \bar{R}_\nu \partial \bar{R}_{\nu'} \partial \bar{R}_{\nu''}} = \frac{\partial \hat{K}_{2,\nu\nu'}}{\partial \bar{R}_{\nu''}}$. With this expansion, the Ehrenfest mean force becomes

$$\bar{F}_\nu = -Tr\left(\hat{\rho}\frac{\partial \hat{H}_e(\hat{R})}{\partial \hat{R}_\nu}\right) \tag{4.19}$$

$$= -Tr\left(\hat{\rho}\frac{\partial}{\partial \bar{R}_\nu}\left(\hat{H}_e(\bar{R}) - \sum_{\nu'} \hat{F}_{\nu'} \, \Delta \hat{R}_{\nu'} + \frac{1}{2}\sum_{\nu'\nu''} \hat{K}_{2,\nu'\nu''} \, \Delta \hat{R}_{\nu'} \Delta \hat{R}_{\nu''} + \dots\right)\right) \tag{4.20}$$

$$= Tr\left(\hat{\rho}\hat{F}_\nu\right) - \sum_{\nu'} Tr\left(\hat{\rho}\hat{K}_{2,\nu\nu'}\Delta \hat{R}_{\nu'}\right) - \frac{1}{2}\sum_{\nu'\nu''} Tr\left(\hat{\rho}\hat{K}_{3,\nu\nu'\nu''}\Delta \hat{R}_{\nu'}\Delta \hat{R}_{\nu''}\right) + \dots \tag{4.21}$$

$$= Tr_e\left(\hat{\rho}_e\hat{F}_\nu\right) - \sum_{\nu'} Tr_e\left(\hat{K}_{2,\nu\nu'}\hat{\mu}_{1,\nu'}\right) - \frac{1}{2}\sum_{\nu'\nu''} Tr_e\left(\hat{K}_{3,\nu\nu'\nu''}\hat{\mu}_{2,\nu'\nu''}\right) + \dots \tag{4.22}$$

where the moment operators are $\hat{\mu}_{1,\nu'} = Tr_I(\Delta \hat{R}_{\nu'}\hat{\rho})$ and $\hat{\mu}_{2,\nu'\nu''}$ $= Tr_I(\Delta \hat{R}_{\nu'} \Delta \hat{R}_{\nu''}\hat{\rho})$. Likewise, other moment operators can be defined for the expansion in momentum $\hat{\lambda}_{1,\nu'} \equiv Tr_I(\Delta \hat{P}_{\nu'}\hat{\rho})$ and $\hat{\lambda}_{2,\nu'\nu''} \equiv Tr_I(\Delta \hat{P}_{\nu'} \Delta \hat{P}_{\nu''}\hat{\rho})$.

No assumption needs to be made on the form of the ionic potential in $\hat{H}_e(\bar{R})$: CEID's expansion is completely general. The first term in Eq. (4.22) is the force in the Ehrenfest approximation, while the other terms represent perturbative corrections depending on the width of the atomic wavepacket. In CEID, the mean atomic force does not depend on one single ionic trajectory as in Ehrenfest, but on the multiple trajectories allowed by the spread of the wavepacket. The moments have a physical interpretation: $\mu_{1,\nu,\alpha\alpha} = \langle\alpha|\hat{\mu}_{1,\nu}|\alpha\rangle/\langle\alpha|\hat{\rho}_e|\alpha\rangle$ is the conditional of the ionic displacement from the mean position if the electrons are in state $|\alpha\rangle$, while $\mu_{2,\nu\nu',\alpha\alpha}$ measures the mean width of the atomic wavepacket.

### 4.3.1 CEID Simulations

To use this scheme in simulations and obtain a closed set of EOM, the expansion must be truncated. The authors of [9, 10] use the order of the moments as a truncation parameter: the sum of the powers of operators $\Delta \hat{R}_\nu$ and $\Delta \hat{P}_\nu$ defines the order. The zeroth moment approximation completely ignores ionic fluctuations and is equivalent to the Ehrenfest approximation. The first moment approximation gives the EOM [9]

$$\bar{F}_\nu = Tr_e\left(\hat{\rho}_e \hat{F}_\nu\right) - \sum_{\nu'} Tr_e\left(\hat{\mu}_{1,\nu'}\hat{K}_{2,\nu\nu'}\right) \tag{4.23}$$

$$\frac{d\hat{\rho}_e}{dt} = \frac{1}{i\hbar}[\hat{H}_e(\bar{R}), \hat{\rho}_e] - \frac{1}{i\hbar}\sum_\nu[\hat{F}_\nu, \hat{\mu}_{1,\nu}] \tag{4.24}$$

$$\frac{d\hat{\mu}_{1,\nu}}{dt} = \frac{1}{i\hbar}[\hat{H}_e(\bar{R}), \hat{\mu}_{1,\nu}] + \frac{\hat{\lambda}_{1,\nu}}{M_\nu} \tag{4.25}$$

$$\frac{d\hat{\lambda}_{1,\nu}}{dt} = \frac{1}{i\hbar}[\hat{H}_e(\bar{R}), \hat{\lambda}_{1,\nu}] + \frac{1}{2}(\Delta \hat{F}_\nu \hat{\rho}_e + \hat{\rho}_e \Delta \hat{F}_\nu) - \frac{1}{2}\sum_{\nu'}(\hat{K}_{2,\nu\nu'}\hat{\mu}_{1,\nu'} + \hat{\mu}_{1,\nu'}\hat{K}_{2,\nu\nu'}) \tag{4.26}$$

where $\Delta \hat{F}_\nu = \hat{F}_\nu - \bar{F}_\nu$, $\Delta \hat{P}_\nu = \hat{P}_\nu - \bar{P}_\nu$ and $\hat{\lambda}_{1,\nu} = Tr_I(\Delta \hat{P}_\nu \hat{\rho})$. The crucial innovation brought forward by CEID is the dispersion in ionic trajectories that causes a spread of forces on the electrons and a spread of electronic trajectories. These electronic trajectories, in turn, generate a spread in ionic forces and trajectories. The term $Tr_e\left(\hat{\mu}_{1,\nu'}\hat{K}_{2,\nu\nu'}\right)$ in Eq. (4.23) describes electronic fluctuations that influence the atomic force $\bar{F}_\nu$. Likewise, in Eq. (4.24), atomic fluctuations $-\frac{1}{i\hbar}\sum_\nu[\hat{F}_\nu, \hat{\mu}_{1,\nu}]$ produce noise in the electronic density $\hat{\rho}_e$. We can see the emergence of correlated fluctuations between individual atoms and individual electrons. The microscopic noise created by an electron-ion collision propagates in the dynamics of both. These fluctuations are simultaneously carried to both subsystems by the same correlation function $\hat{\mu}_{1,\nu}$. This microscopic noise is the source of mutual heat exchange missing in the Ehrenfest approximation.

The authors in [9] simulate an atomic wire with one atom in the center allowed to move. They aim to let an electronic current flow and observe the effect of the current on the central atom. They employ CEID with a truncation at the first moment. Their attempt is successful in observing a significant increase in energy of the moving ion for increasing biases, but fails to observe a corresponding reduction in current. A high ionic energy would cause an increased ionic scattering that is expected to increase the resistance and lower the current.

In [18], the authors observe that there is no difference in the current time evolution for simulations based on the Ehrenfest approximation and on first moment CEID, as shown in Fig. 4.4. First moment CEID improves on the Ehrenfest approximation by

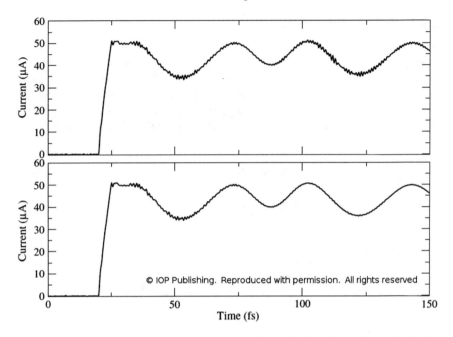

**Fig. 4.4** Current dynamics in a nanowire where a central atom is allowed to oscillate with the first moment CEID (top curve) and the Ehrenfest approximation (bottom curve), [18]

capturing the current induced ionic heating, but it is still not able to capture the ionic feedback on the current.

The key quantity for capturing this missing effect is the atomic width $\hat{\mu}_{2,\nu\nu'}$, which affects the electron dynamics and acts as a current dampening mechanism. The first moment truncation ignores it. To reintroduce it, CEID must be truncated at least at the second order and this implementation is described in [10]. As shown in Fig. 4.5, the authors attempt a new set of simulations and do observe the expected inelastic decrease in the current. They also see a clear signature of the inelastic excitations of phonons in the derivative of the differential conductance by noting a shoulder at the voltage where the inelastic process is activated.

The electronic current in the finite nanowires above is determined by a transient phase of charge imbalance. The current flow is temporary and systems can only reach temporary steady states. In [19], the authors developed a time dependent open boundary (OB) formalism for electron injection and extraction, which guarantees a steady flow of electrons. Because of the relevance of the OB mechanism to this present work, a schematic derivation of it can be found in Appendix B. An application of the OB scheme to second moment CEID [19] shows the expected current induced ionic heating together with the insurgence of a steady state current, as can be seen in Fig. 4.6. The characteristic frequency of the phonon $\hbar\omega$ represent a threshold for inelastic excitations. At low voltages, below the threshold, vibrational excitations are

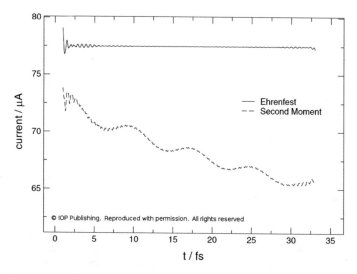

**Fig. 4.5** Comparison of the time evolution of the current in an atomic wire with a moving atom for simulations using the Ehrenfest approximation and the second moment CEID [10]. The Ehrenfest approximation lacks the feedback of atomic fluctuations on the electron dynamics, thus it cannot capture the ionic induced reduction in current

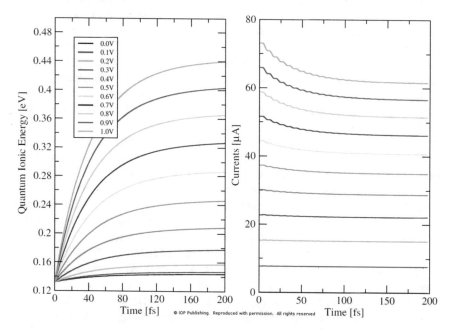

**Fig. 4.6** Vibrational energy of a single vibrating atom in a nanowire with a current induced by OB and the corresponding current at several voltages, [19]

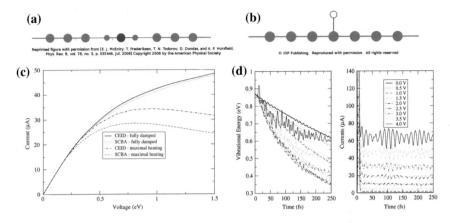

**Fig. 4.7** Second moment CEID simulations of nanowires under current. A trimer (panel (**a**)) displays negative differential resistance at high voltages (panel (**b**)) [20]. An atom is weakly coupled to a perfect chain (panel (**c**)) and cools down with a current passing along the wire (panel (**d**)) [21]

forbidden and no heating is observed, as the inelastic current is suppressed. As the voltage is increased the steady state current increases and the ion heats up.

CEID has been applied to a number of systems and it allowed the observation of a number of non-adiabatic effects at the nanoscale. In [20], the authors consider a trimer weakly coupled to the electrodes and with only the central atom free to oscillate, as in Fig. 4.7a. The system is resonant and shows peculiar features when compared with an equivalent ballistic atomic chain: a significantly higher phonon occupancy for high voltages and an increase in the time taken to equilibrate. Two cases are explored: the fully damped one, where the phonon mode is forced to stay in its ground state and the maximal heating case, where the atom is free to be excited. In Fig. 4.7c we see that, at high voltages, undamped CEID displays a negative differential resistance.

An adatom system made of an infinite chain with a single light atom attached and free to move is examined in [21] and shown in Fig. 4.7b. This system is antiresonant and displays a current-assisted cooling, as in Fig. 4.7d. The oscillator starts from an excited state, and an electronic current helps to stabilize the system by cooling down the atom. For increasingly high biases, the initial rate of cooling of the oscillator increases, identifying the current as the cooling down mechanism.

Despite its successes in describing non-adiabatic phenomena, the CEID method presents some significant shortcomings. One is the choice of a truncation strategy, i.e. how to cut the Taylor expansion in the moments and derive a closed set of equations. The freedom of choice for the atomic potential makes CEID very general, but at the same time, makes a general physical truncation very hard to figure out. A truncation strategy is proposed in [10] and the subject is reported as under ongoing work.

Probably, CEID's main limitation is its scaling with the number of moving atoms. In second moment CEID, the numerical scaling of the integration of the equations of motion is quadratic with the number of moving atoms. This makes simulations of anything but a few atomic degrees of freedom an insurmountable task. The ECEID method that we introduce in Chap. 5 and is the subject of this thesis goes beyond CEID's limitation by focusing on the case of harmonic potentials and proposing a clear truncation strategy.

## 4.4  The Bonca-Trugman Method

Another method that solves the electron-phonon problem on a model system was developed in [22]. It is based on mapping the full electron-phonon many-body problem into a one-body problem by providing a cutoff in phonon space. The size of the variational space can be tuned so that the solution converges to the exact one.

The authors consider a single electron in the presence of a phonon and use this second-quantization model Hamiltonian

$$\hat{H} = \hat{H}_{el} + \hat{H}_{ph} + \hat{H}_{el-ph} \tag{4.27}$$

$$\hat{H}_{el} = \sum_j \epsilon_j \hat{c}_j^\dagger \hat{c}_j - \sum_{j,k} t_{j,k} \, \hat{c}_j^\dagger \hat{c}_k \tag{4.28}$$

$$\hat{H}_{ph} = \sum_m \omega_m \hat{a}_m^\dagger \hat{a}_m \tag{4.29}$$

$$\hat{H}_{el-ph} = - \sum_{j,k,m} \gamma_{j,k,m} \, \hat{c}_j^\dagger \hat{c}_k (\hat{a}_m^\dagger + \hat{a}_m) \tag{4.30}$$

where the electronic one, $\hat{H}_{el}$, has an onsite energy $\epsilon_j$ for site $j$ and hopping $t_{j,k}$ between sites $j$ and $k$ and the phonon one, $\hat{H}_{ph}$, includes $m$ modes with frequencies $\omega_m$. The electron-phonon coupling $\hat{H}_{el-ph}$ can include both diagonal and off-diagonal contributions on the electronic basis.

To illustrate the method, the authors choose one phonon mode coupled to electronic site zero and a maximum phonon occupancy of 2. An electron enters the system as a plane wave on the left lead. The electron can be transmitted or backscattered elastically or inelastically. Their method consists of pruning any site that only contains an outgoing wave. This procedure eliminates some states from the problem while changing the onsite energy of some states to complex, as is schematically depicted in Fig. 4.8a. In in [22], the authors proceed solving the problem for a number of different parameters and analyze processess of inelastic resonant tunnelling.

Among several applications, the Bonca-Trugman method was used to investigate polaron formation in a one-dimesional chain with an impurity allowed to vibrate [23], a problem that was studied earlier in [24]. There, the system is made of 5 electronic sites and 1 phonon coupled to the central site. A master equation approach is used to couple this system to a bath of vibrations kept at a constant temperature. The authors simulate an electron localized on the leftmost site and a phonon at its ground state and compare the quantum method with Ehrennfest dynamics. The population dynamics in real time of the electronic sites population shows very fast oscillations

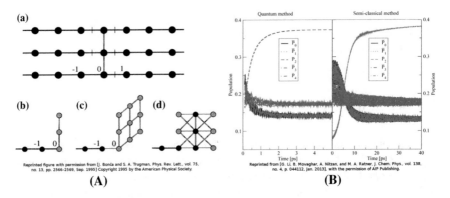

**Fig. 4.8** In panel (**A**), graphical representation of how the Bonca-Trugman method works [22]. Each dot represents a state in the coupled electron-phonon system, where the horizontal axis represents electronic sites and the vertical axis the phonon occupation. Lines represent nonzero coupling between states. In **a**, a system where electronic site zero is coupled to a phonon mode, which is restricted to occupations 0, 1 and 2. **b** contains the system after the pruning procedure in the text, where the light grey dots present complex onsite energies. **c** the same pruned system, now containing 2 phonons. **d** a system where the phonon is coupled also to electronic sites −1 and 1. In panel (**B**) average population of the 5 sites in a chain coupled to an oscillator on the central site and comparison of the Bonca-Trugman method with Ehrenfest dynamics [23]. The details about the model are reported in the article

due to coherent reflections within such a small system. To see a long time trend, a coarse-grained time evolution of the population is needed and is shown in Fig. 4.8b.

Both methods show a qualitatively similar behavior, with the electron localizing on the central site, with the quantum method doing so about one order of magnitude faster. The use of a mean field approximation in Ehrenfest dynamics makes the potential surface less binding than in the quantum case, slowing down localization. An increasing hopping in the chain makes the electron more delocalized and the central site population smaller, while a strong electron-phonon coupling forms a polaron with a larger population on the central site.

In a recent article [25], the authors take a similar model Hamiltonian and apply a Lanczos-based diagonalization that selects states with different phonon occupations around the electron to analyze thermalization. They show that non-equilibrium systems evolve towards thermal states by tracking the occupations of fermionic momenta starting at different conditions. The sheer size of the space of the fully coupled system is a limit for their simulations that include a single electron on 16 sites and 1 phonon.

The aim of this thesis is to simulate systems with a number of electrons and phonons on femtosecond to picosecond timescales and to keep track of their simultaneous evolution in out of equilibrium situations. The methods based on the 2TM from Chap. 3 are overly simplified to fulfill our aim because they don't consider electrons as quantum particles, while the non-adiabatic methods described in this chapter either cannot capture the correct heat exchange or are too computationally expensive. Despite the relevant successes that these methods had in the fields that they were designed for, none is perfectly suitable for the objectives in this thesis. The desire to achieve our aim is the motivation for our development of a new method that we present in the next chapter.

# References

1. Stefanucci, G. and R. van Leeuwen. 2013. *Nonequilibrium many-body theory of quantum systems: a modern introduction*. Cambridge University Press. ISBN 9780521766173.
2. Galperin, M., M.A. Ratner, and A. Nitzan. 2004. https://doi.org/10.1063/1.1814076 Inelastic electron tunneling spectroscopy in molecular junctions: peaks and dips. *The Journal of Chemical Physics* 121 (23): 11 965–79. https://doi.org/10.1063/1.1814076.
3. Frederiksen, T., M. Brandbyge, N. Lorente, and A.-P. Jauho. 2004. https://doi.org/10.1103/PhysRevLett.93.256601 Inelastic scattering and local heating in atomic gold wires. *Physical Review Letters* 93 (25): 256601. https://doi.org/10.1103/PhysRevLett.93.256601.
4. Zhu, Y., J. Maciejko, T. Ji, H. Guo, and J. Wang. 2005. https://doi.org/10.1103/PhysRevB.71.075317 Time-dependent quantum transport: Direct analysis in the time domain. *Physical Review B* 71 (7): 075317. https://doi.org/10.1103/PhysRevB.71.075317.
5. Stella, L., M. Meister, A.J. Fisher, and A.P. Horsfield. 2007. https://doi.org/10.1063/1.2801537 Robust nonadiabatic molecular dynamics for metals and insulators. *The Journal of Chemical Physics* 127 (21): 214104. https://doi.org/10.1063/1.2801537.
6. Stella, L., R.P. Miranda, A.P. Horsfield, and A.J. Fisher. 2011. https://doi.org/10.1063/1.3589165 Analog of Rabi oscillations in resonant electron-ion systems. *The Journal of Chemical Physics* 134 (19): 194105. https://doi.org/10.1063/1.3589165.
7. Horsfield, A.P., D.R. Bowler, H. Ness, C.G. Sánchez, T.N. Todorov, and A.J. Fisher. 2006. https://doi.org/10.1088/0034-4885/69/4/R05 The transfer of energy between electrons and ions in solids. *Reports on Progress in Physics* 69 (4): 1195–1234. https://doi.org/10.1088/0034-4885/69/4/R05.
8. Horsfield, A.P., D.R. Bowler, A.J. Fisher, T.N. Todorov, and M.J. Montgomery. 2004. https://doi.org/10.1088/0953-8984/16/21/010 Power dissipation in nanoscale conductors: classical, semi-classical and quantum dynamics. *Journal of Physics: Condensed Matter* 16 (21): 3609–3622. https://doi.org/10.1088/0953-8984/16/21/010.
9. Horsfield, A.P., D.R. Bowler, A.J. Fisher, T.N. Todorov, C.G. Sánchez. 2004. https://doi.org/10.1088/0953-8984/16/46/012 Beyond Ehrenfest: correlated non-adiabatic molecular dynamics. *Journal of Physics: Condensed Matter* 16 (46): 8251–8266. https://doi.org/10.1088/0953-8984/16/46/012.
10. Horsfield, A.P., D.R. Bowler, A.J. Fisher, T.N. Todorov, and C.G. Sanchez. 2005. https://doi.org/10.1088/0953-8984/17/30/006 Correlated electron-ion dynamics: the excitation of atomic motion by energetic electrons. *Journal of Physics: Condensed Matter* 17 (30): 4793–4812. https://doi.org/10.1088/0953-8984/17/30/006.
11. le Page, J., D.R. Mason, C.P. Race, and W.M.C. Foulkes. 2009. https://doi.org/10.1088/1367-2630/11/1/013004 How good is damped molecular dynamics as a method to simulate radiation damage in metals? *New Journal of Physics* 11 (1): 013004. https://doi.org/10.1088/1367-2630/11/1/013004.
12. Finnis, M.W., P. Agnew, and A.J.E. Foreman. 1991. https://doi.org/10.1103/PhysRevB.44.567 Thermal excitation of electrons in energetic displacement cascades. *Physical Review B* 44 (2): 567–574. https://doi.org/10.1103/PhysRevB.44.567.
13. Nordlund, K., M. Ghaly, and R.S. Averback. 1998. https://doi.org/10.1063/1.366821 Mechanisms of ion beam mixing in metals and semiconductors. *Journal of Applied Physics* 83 (3): 1238. https://doi.org/10.1063/1.366821.

14. Duffy, D.M., and A.M. Rutherford. 2007. https://doi.org/10.1088/0953-8984/19/1/016207
    Including the effects of electronic stopping and electron-ion interactions in radiation damage
    simulations. *Journal of Physics: Condensed Matter* 19 (1): 016207. https://doi.org/10.1088/
    0953-8984/19/1/016207.
15. Correa, A.A., J. Kohanoff, E. Artacho, D. Sánchez-Portal, and A. Caro. 2012. https://doi.org/
    10.1103/PhysRevLett.108.213201 Nonadiabatic forces in ion-solid interactions: the initial
    stages of radiation damage. *Physical Review Letters* 108 (21): 213201. https://doi.org/10.
    1103/PhysRevLett.108.213201.
16. Zeb, M.A., J. Kohanoff, D. Sánchez-Portal, A. Arnau, J.I. Juaristi, and E. Artacho. 2012.
    https://doi.org/10.1103/PhysRevLett.108.225504 Electronic stopping power in gold: the role
    of d electrons and the H/He anomaly. *Physical Review Letters* 108 (22): 225504. https://doi.
    org/10.1103/PhysRevLett.108.225504.
17. Ullah, R., F. Corsetti, D. Sánchez-Portal, and E. Artacho. 2015. https://doi.org/10.1103/
    PhysRevB.91.125203 Electronic stopping power in a narrow band gap semiconductor from
    first principles. *Physical Review B* 91 (12): 125203. https://doi.org/10.1103/PhysRevB.91.
    125203.
18. Bowler, D.R., A.P. Horsfield, C.G. Sánchez, and T.N. Todorov. 2005. https://doi.org/10.1088/
    0953-8984/17/25/024 Correlated electron-ion dynamics with open boundaries: formalism.
    *Journal of Physics. Condensed Matter: An Institute of Physics Journal* 17 (25): 3985–95.
    https://doi.org/10.1088/0953-8984/17/25/024.
19. McEniry, E.J., D.R. Bowler, D. Dundas, A.P. Horsfield, C.G. Sánchez, and T.N. Todorov.
    2007. https://doi.org/10.1088/0953-8984/19/19/196201 Dynamical simulation of inelastic
    quantum transport. *Journal of Physics: Condensed Matter* 19 (19): 196201. https://doi.org/
    10.1088/0953-8984/19/19/196201.
20. McEniry, E., T. Frederiksen, T. Todorov, D. Dundas, and A. Horsfield. 2008. https://doi.
    org/10.1103/PhysRevB.78.035446 Inelastic quantum transport in nanostructures: The self-
    consistent Born approximation and correlated electron-ion dynamics. *Physical Review B* 78
    (3): 035446. https://doi.org/10.1103/PhysRevB.78.035446.
21. McEniry, E.J., T.N. Todorov, and D. Dundas. 2009. https://doi.org/10.1088/0953-8984/21/
    19/195304 Current-assisted cooling in atomic wires. *Journal of Physics: Condensed Matter*
    21 (19): 195304. https://doi.org/10.1088/0953-8984/21/19/195304.
22. Bonča, J., and S.A. Trugman. 1995. https://doi.org/10.1103/PhysRevLett.75.2566 Effect of
    inelastic processes on tunneling. *Physical Review Letters* 75 (13): 2566–2569. https://doi.org/
    10.1103/PhysRevLett.75.2566.
23. G. Li, B. Movaghar, A. Nitzan, and M.A. Ratner. 2013. https://doi.org/10.1063/1.4776230
    Polaron formation: Ehrenfest dynamics vs. exact results. *The Journal of Chemical Physics*
    138 (4): 044112. https://doi.org/10.1063/1.4776230.
24. Ness, H., S.A. Shevlin, and A.J. Fisher. 2001. https://doi.org/10.1103/PhysRevB.63.125422
    Coherent electron-phonon coupling and polaronlike transport in molecular wires. *Physical
    Review B* 63 (12): 125422. https://doi.org/10.1103/PhysRevB.63.125422.
25. Kogoj, J., L. Vidmar, M. Mierzejewski, S.A. Trugman, and J. Bonča. 2016. https://doi.org/
    10.1103/PhysRevB.94.014304 Thermalization after photoexcitation from the perspective of
    optical spectroscopy. *Physical Review B - Condensed Matter and Materials Physics* 94 (1):
    014304. https://doi.org/10.1103/PhysRevB.94.014304.

# Chapter 5
# The ECEID Method

In this chapter we present the derivation of the method that we developed and used in our simulations: Effective Correlated Electron-Ion Dynamics (ECEID). We start from the model Hamiltonian and, after some approximations, we obtain a set of equations of motion (EOM) for electronic and vibrational quantities. We also discuss total energy conservation, the Open-Boundaries implementation and the Many-Body to One-Body projection of the EOM. We then outline the implementation of the method in a code and perform scaling tests with a varying number of oscillators and electronic sites. In Chap. 9 will present a recent and more general reformulation of the method.

## 5.1 The Model

We start from model Hamiltonian

$$\hat{H} = \underbrace{\hat{H}_e + \sum_{\nu=1}^{N_o} \left( \frac{\hat{P}_\nu^2}{2M_\nu} + \frac{1}{2} K_\nu \hat{X}_\nu^2 \right) - \sum_{\nu=1}^{N_o} \hat{F}_\nu \hat{X}_\nu}_{\hat{H}_0} \qquad (5.1)$$

where $\hat{H}_e$ is a general many-electron Hamiltonian without vibrations. $\hat{X}_\nu$ and $\hat{P}_\nu$ are displacement and canonical momentum operators for oscillator $\nu$, with mass $M_\nu$ and spring constant $K_\nu$, coupled linearly to the electrons via the electronic operator $\hat{F}_\nu$. $N_o$ is the number of harmonic vibrational degrees of freedom (DOF). Any harmonic Hamiltonian in the vibrational DOF can be written in this form through a change of generalized coordinates. The unperturbed Hamiltonian $\hat{H}_0$ merges $\hat{H}_e$ with the harmonic oscillator Hamiltonian, excluding the mixed electron-oscillator coupling

© Springer International Publishing AG, part of Springer Nature 2018
V. Rizzi, *Real-Time Quantum Dynamics of Electron-Phonon Systems*,
Springer Theses, https://doi.org/10.1007/978-3-319-96280-1_5

terms. In the following derivation we will use the fact that oscillator operators, such as $\hat{X}_\nu$ and $\hat{P}_\nu$, trivially commute with electronic ones such as $\hat{H}_e$ and $\hat{F}_\nu$.

The full system density matrix (DM) evolves according to the Liouville equation

$$\dot{\hat{\rho}}(t) = \frac{1}{i\hbar}\,[\hat{H}, \hat{\rho}(t)]. \tag{5.2}$$

By taking the following time derivative

$$\frac{d}{dt}(e^{\frac{i}{\hbar}\hat{H}_0 t}\hat{\rho}(t)e^{-\frac{i}{\hbar}\hat{H}_0 t}) = \frac{1}{i\hbar}e^{\frac{i}{\hbar}\hat{H}_0 t}\sum_{\nu=1}^{N_0}[-\hat{F}_\nu\hat{X}_\nu, \hat{\rho}(t)]e^{-\frac{i}{\hbar}\hat{H}_0 t} \tag{5.3}$$

and integrating it in time, the full DM can be written exactly as

$$\hat{\rho}(t) = e^{-\frac{i}{\hbar}\hat{H}_0 t}\hat{\rho}(0)e^{\frac{i}{\hbar}\hat{H}_0 t} - \frac{1}{i\hbar}\sum_{\nu=1}^{N_0}\int_0^t e^{\frac{i}{\hbar}\hat{H}_0(\tau-t)}[\hat{F}_\nu\hat{X}_\nu, \hat{\rho}(\tau)]\,e^{-\frac{i}{\hbar}\hat{H}_0(\tau-t)}d\tau. \tag{5.4}$$

The electronic DM $\hat{\rho}_e(t) = \mathrm{Tr}_0(\hat{\rho}(t))$ obeys the effective Liouville equation [1]

$$\dot{\hat{\rho}}_e(t) = \frac{1}{i\hbar}\,[\hat{H}_e, \hat{\rho}_e(t)] - \frac{1}{i\hbar}\sum_{\nu=1}^{N_0}[\hat{F}_\nu, \hat{\mu}_\nu(t)] \tag{5.5}$$

that we obtain by tracing Eq. (5.2) over the oscillator DOF. We define the electronic operator

$$\hat{\mu}_\nu(t) = \mathrm{Tr}_0(\hat{X}_\nu\hat{\rho}(t)) \tag{5.6}$$

that keeps track of the correlation between electrons and phonons and feeds it back to $\hat{\rho}_e(t)$.

While we describe the electronic dynamics with the electronic DM, the dynamical quantity that we employ to follow the time evolution of the phonons is the mean oscillator occupation $N_\nu(t) = \mathrm{Tr}(\hat{N}_\nu\hat{\rho}(t))$. It enters the oscillator Hamiltonian as $(\hat{N}_\nu + \frac{1}{2})\hbar\omega_\nu = \frac{\hat{P}_\nu^2}{2M_\nu} + \frac{1}{2}K_\nu\hat{X}_\nu^2$, with the characteristic oscillator frequency $\omega_\nu = \sqrt{K_\nu/M_\nu}$. In second quantization $\hat{N}_\nu = \hat{a}_\nu^\dagger\hat{a}_\nu$, where $\hat{a}_\nu^\dagger$ ($\hat{a}_\nu$) are the creation (annihilation) operators for oscillator $\nu$ satisfying the canonical relation $[\hat{a}_\nu, \hat{a}_{\nu'}^\dagger] = \delta_{\nu\nu'}$. The canonical displacement is

$$\hat{X}_\nu = \sqrt{\hbar/(2M_\nu\omega_\nu)}(\hat{a}_\nu + \hat{a}_\nu^\dagger) \tag{5.7}$$

and the momentum

$$\hat{P}_\nu = i\sqrt{\hbar M_\nu\omega_\nu/2}(\hat{a}_\nu^\dagger - \hat{a}_\nu). \tag{5.8}$$

The time derivative of $N_\nu(t)$ can be written as

$$\dot{N}_v(t) = \mathrm{Tr}\left(\hat{N}_v \dot{\hat{\rho}}(t)\right) = -\frac{1}{i\hbar}\mathrm{Tr}\left([\hat{N}_v, \hat{F}_v \hat{X}_v]\hat{\rho}(t)\right) \tag{5.9}$$

where we used Eq. (5.5) and rearranged the operators in the trace, keeping in mind that $[\hat{N}_v, \hat{H}_0] = 0$.[1] By using the canonical commutation relation and the second quantization form of the oscillator operators,[2] we can write Eq. (5.9) in a more compact way as

$$\dot{N}_v(t) = \frac{1}{\hbar M_v \omega_v}\mathrm{Tr}\left(\hat{F}_v \hat{P}_v \hat{\rho}(t)\right) = \frac{1}{\hbar M_v \omega_v}\mathrm{Tr}_e\left(\hat{F}_v \hat{\lambda}_v(t)\right), \tag{5.10}$$

where we defined

$$\hat{\lambda}_v(t) = \mathrm{Tr}_0(\hat{P}_v \hat{\rho}(t)). \tag{5.11}$$

The dynamics of $\hat{\rho}_e(t)$ and $N_v(t)$ is controlled by electronic operators $\hat{\mu}_v(t)$ and $\hat{\lambda}_v(t)$ that are the crucial correlation operators linking electrons and oscillators. In the next section, we provide exact expressions for these quantities, paving the way for the approximations that follow.

## 5.2 An Exact Form of $\hat{\mu}_v(t)$ and $\hat{\lambda}_v(t)$

We introduce the notation

$$\hat{Q}^t = e^{\frac{i}{\hbar}\hat{H}_0 t}\hat{Q}e^{-\frac{i}{\hbar}\hat{H}_0 t} \tag{5.12}$$

for a generic operator $\hat{Q}$. This short notation is convenient in the following derivations where the implicit time dependency of many operators can be written in such a concise way.

We start by inserting Eq. (5.4) into the definition of $\hat{\mu}_v(t)$

$$\hat{\mu}_v(t) = -\frac{1}{i\hbar}\mathrm{Tr}_0\left(\hat{X}_v \sum_{v'=1}^{N_0}\int_0^t [\hat{F}_{v'}^{\tau-t}\hat{X}_{v'}^{\tau-t}, \hat{\rho}^{\tau-t}(\tau)]\, d\tau\right). \tag{5.13}$$

Here we assume for simplicity that the unperturbed motion described by $\hat{\rho}^{-t}(0)$ in (5.4) does not contribute to the motion of the oscillator centroids and to the dynamics of $\hat{\mu}_v(t)$.

We expand the commutator and permute the operators within the oscillator trace in Eq. (5.13) to obtain

---

[1] Notice that this commutator would *not* be zero if there was an anharmonic contribution in $\hat{H}_0$ for example.

[2] It is easy to see that $[\hat{N}_v, \hat{X}_v] = \sqrt{\hbar/(2M_v\omega_v)}[\hat{a}_v^\dagger \hat{a}_v, \hat{a}_v + \hat{a}_v^\dagger] = -(i/(M_v\omega_v))\hat{P}_v$.

$$\hat{\mu}_\nu(t) = -\frac{1}{i\hbar} \text{Tr}_\text{o} \sum_{\nu'=1}^{N_\text{o}} \left( \int_0^t \hat{F}_{\nu'}^{\tau-t} \hat{X}_\nu \hat{X}_{\nu'}^{\tau-t} \hat{\rho}^{\tau-t}(\tau)\,d\tau - \int_0^t \hat{\rho}^{\tau-t}(\tau)\hat{X}_{\nu'}^{\tau-t} \hat{X}_\nu \hat{F}_{\nu'}^{\tau-t}\,d\tau \right).$$

$$(5.14)$$

By applying the decomposition

$$\hat{A}\hat{B} = \frac{1}{2}\{\hat{A}, \hat{B}\} + \frac{1}{2}[\hat{A}, \hat{B}] \tag{5.15}$$

to both $\hat{X}_\nu \hat{X}_{\nu'}^{\tau-t}$ and $\hat{X}_{\nu'}^{\tau-t} \hat{X}_\nu$, Eq. (5.14) can be written as

$$\hat{\mu}_\nu(t) = -\frac{1}{2i\hbar} \text{Tr}_\text{o} \sum_{\nu'=1}^{N_\text{o}} \left( \int_0^t [\hat{F}_{\nu'}^{\tau-t}, \hat{\rho}^{\tau-t}(\tau)]\{\hat{X}_\nu, \hat{X}_{\nu'}^{\tau-t}\}\,d\tau \right)$$

$$-\frac{1}{2i\hbar} \text{Tr}_\text{o} \sum_{\nu'=1}^{N_\text{o}} \left( \int_0^t \{\hat{F}_\nu^{\tau-t}, \hat{\rho}^{\tau-t}(\tau)\}[\hat{X}_\nu, \hat{X}_{\nu'}^{\tau-t}]\,d\tau \right). \tag{5.16}$$

The second derivative with respect to time of $\hat{X}_\nu^{\tau-t}$ satisfies the usual harmonic oscillator differential equation[3]

$$\ddot{\hat{X}}_\nu^{\tau-t} = -\omega_\nu^2 \hat{X}_\nu^{\tau-t}, \tag{5.17}$$

as can be verified by using the canonical position-momentum commutation relation $[\hat{X}_\nu, \hat{P}_{\nu'}] = i\hbar\delta_{\nu\nu'}$. The solution of (5.17), with initial conditions $\hat{X}_\nu^0 = \hat{X}_\nu$ and $\dot{\hat{X}}_\nu^0 = \hat{P}_\nu/M_\nu$, is

$$\hat{X}_\nu^{\tau-t} = \hat{X}_\nu \cos\omega_\nu(\tau - t) + \frac{\hat{P}_\nu}{M_\nu\omega_\nu} \sin\omega_\nu(\tau - t), \tag{5.18}$$

which can be rewritten in second quantization as

$$\hat{X}_\nu^{\tau-t} = \sqrt{\frac{\hbar}{2M_\nu\omega_\nu}} (\hat{a}_\nu^\dagger e^{i\omega_\nu(\tau-t)} + \hat{a}_\nu e^{-i\omega_\nu(\tau-t)}), \tag{5.19}$$

where the canonical position and momentum operators are (5.7) and (5.8).

Now, we can insert

$$\sum_{\nu'=1}^{N_\text{o}}[\hat{X}_\nu, \hat{X}_{\nu'}^{\tau-t}] = \frac{i\hbar}{M_\nu\omega_\nu} \sin\omega_\nu(\tau - t). \tag{5.20}$$

---

[3]To prove it, we write down this time derivative $\dot{\hat{X}}_\nu^{\tau-t} = e^{\frac{i}{\hbar}\hat{H}_0(\tau-t)} \frac{[\hat{X}_\nu, \hat{H}_0]}{i\hbar} e^{-\frac{i}{\hbar}\hat{H}_0(\tau-t)}$. By solving the commutator and rearranging terms, we obtain $\dot{\hat{X}}_\nu^{\tau-t} = \frac{\hat{P}_\nu^{\tau-t}}{M_\nu}$. Analogously, after taking another time derivative, we have Eq. (5.17).

into the second term of Eq. (5.16) and obtain

$$\hat{\mu}_\nu(t) = -\frac{1}{2i\hbar} \, \mathrm{Tr}_o \sum_{\nu'=1}^{N_o} \left( \int_0^t [\hat{F}_{\nu'}^{\tau-t}, \hat{\rho}^{\tau-t}(\tau)]\{\hat{X}_\nu, \hat{X}_{\nu'}^{\tau-t}\} \, d\tau \right)$$
$$- \frac{1}{2M_\nu\omega_\nu} \, \mathrm{Tr}_o \left( \int_0^t \{\hat{F}_\nu^{\tau-t}, \hat{\rho}^{\tau-t}(\tau)\} \sin\omega_\nu(\tau-t) \, d\tau \right), \qquad (5.21)$$

which is *exact*.

We can apply a similar strategy to $\hat{\lambda}_\nu(t)$, and insert the full DM (5.4) into its definition to get

$$\hat{\lambda}_\nu(t) = -\frac{1}{i\hbar} \mathrm{Tr}_o\left( \hat{P}_\nu \int_0^t \sum_{\nu'=1}^{N_o} [\hat{F}_{\nu'}^{\tau-t} \hat{X}_{\nu'}^{\tau-t}, \hat{\rho}^{\tau-t}(\tau)] \, d\tau \right) \qquad (5.22)$$

With analogous steps as above, we have

$$\hat{\lambda}_\nu(t) = -\frac{1}{2i\hbar} \mathrm{Tr}_o \sum_{\nu'=1}^{N_o} \left( \int_0^t [\hat{F}_{\nu'}^{\tau-t}, \hat{\rho}^{\tau-t}(\tau)]\{\hat{P}_\nu, \hat{X}_{\nu'}^{\tau-t}\} \, d\tau \right)$$
$$- \frac{1}{2i\hbar} \mathrm{Tr}_o \sum_{\nu'=1}^{N_o} \left( \int_0^t \{\hat{F}_{\nu'}^{\tau-t}, \hat{\rho}^{\tau-t}(\tau)\}[\hat{P}_\nu, \hat{X}_{\nu'}^{\tau-t}] \, d\tau \right) \qquad (5.23)$$

that, with the use of Eq. (5.15) and

$$\sum_{\nu'=1}^{N_o} [\hat{P}_\nu, \hat{X}_{\nu'}^{\tau-t}] = -i\hbar \cos\omega_\nu(\tau-t)\delta_{\nu\nu'} \qquad (5.24)$$

acquires this exact form

$$\hat{\lambda}_\nu(t) = -\frac{1}{2i\hbar} \mathrm{Tr}_o \sum_{\nu'=1}^{N_o} \left( \int_0^t [\hat{F}_{\nu'}^{\tau-t}, \hat{\rho}^{\tau-t}(\tau)]\{\hat{P}_\nu, \hat{X}_{\nu'}^{\tau-t}\} \, d\tau \right)$$
$$+ \frac{1}{2} \mathrm{Tr}_o\left( \int_0^t \{\hat{F}_\nu^{\tau-t}, \hat{\rho}^{\tau-t}(\tau)\} \cos\omega_\nu(\tau-t) \, d\tau \right). \qquad (5.25)$$

These exact forms for $\hat{\mu}_\nu(t)$ and $\hat{\lambda}_\nu(t)$ could already be plugged into the EOM of $\hat{\rho}_e$ and $N_\nu$. The resulting set of equations would be formally exact but it could not be time evolved in simulations yet, due to the presence of expensive time integrals and the unknown anticommutator factors that they contain. The terms involving $\{\hat{X}_\nu, \hat{X}_{\nu'}^{\tau-t}\}$ and $\{\hat{P}_\nu, \hat{X}_{\nu'}^{\tau-t}\}$ are challenging as they have no exact closed form. They require approximations.

## 5.3  The Approximations

To make the set of equations tractable, we devise an approximate scheme to make the time evolution of the equations suitable for simulation. This scheme has the difficult task to make the equations lose enough complexity for the problem to be manageable and, at the same time, to keep enough information for the relevant physics of the electron–phonon processes to be preserved.

We make three approximations. First, in Eqs. (5.21) and (5.25) (but not earlier), we decompose the full system DM into $\hat{\rho}(\tau) \simeq \hat{\rho}_e(\tau)\hat{\rho}_o(\tau)$. This retains electron–phonon correlation exactly to lowest order in the coupling $\hat{F}_\nu$, and approximately to higher order, in analogy to the self-consistent Born approximation [2]. If this decomposition were applied before (e.g. in the definition of $\hat{\mu}_\nu$, before the insertion of (5.4)), the resulting loss of correlation would be much more significant and it would make the method equivalent to Ehrenfest dynamics.

Second, after taking oscillator traces, we retain only terms diagonal in $\nu$, suppressing phonon-phonon correlation.[4] Indirect interactions between phonons can still take place, with the mediation of the electronic subsystem.[5] Third and last, we neglect terms of the form[6] $\mathrm{Tr}_o(\hat{a}_\nu\hat{a}_\nu\hat{\rho}_o^{\tau-t}(\tau))$, $\mathrm{Tr}_o(\hat{a}_\nu^\dagger\hat{a}_\nu^\dagger\hat{\rho}_o^{\tau-t}(\tau))$, retaining only single-phonon processes and excluding anharmonicity. This approximation can be understood as a restriction to a low electron–phonon coupling regime, where high order processes such as double (de)excitations are less relevant.

We single out part of the first term of Eq. (5.21) and apply the above approximations, getting

$$
\mathrm{Tr}_o \sum_{\nu'=1}^{N_o} \left( \{\hat{X}_\nu, \hat{X}_{\nu'}^{\tau-t}\} \hat{\rho}^{\tau-t}(\tau) \right) \simeq \mathrm{Tr}_o \sum_{\nu'=1}^{N_o} \left( \{\hat{X}_\nu, \hat{X}_{\nu'}^{\tau-t}\} \hat{\rho}_o^{\tau-t}(\tau) \right) \hat{\rho}_e^{\tau-t}(\tau)
$$

$$
\simeq \frac{\hbar}{2M_\nu\omega_\nu} \mathrm{Tr}_o \left( \{\hat{a}_\nu^\dagger + \hat{a}_\nu, \hat{a}_\nu^\dagger e^{i\omega_\nu(\tau-t)} + \hat{a}_\nu e^{-i\omega_\nu(\tau-t)}\} \hat{\rho}_o^{\tau-t}(\tau) \right) \hat{\rho}_e^{\tau-t}(\tau)
$$

$$
\simeq \frac{\hbar}{M_\nu\omega_\nu} (2N_\nu(\tau) + 1) \cos\omega_\nu(\tau - t)\, \hat{\rho}_e^{\tau-t}(\tau) \tag{5.26}
$$

where in the first line we have used the decomposition of the DM, in the second line Eqs. (5.19) and (5.7) and the suppression of correlation between different oscillators and in the third line we ignored double (de)excitations.[7] With this approximated

---

[4]Even if there is no explicit phonon-phonon interaction in the model Hamiltonian (5.1), phonon interaction appears in the dynamics of electronic operators $\hat{\mu}_\nu(t)$ and $\hat{\lambda}_\nu(t)$.

[5]For example, the dynamics of phonon 1 can influence the dynamics of the electrons, that, in turn, determine the time evolution of phonon 2.

[6]In Appendix D, we show that it is possible not to invoke this approximation, within the framework of ECEID xp from Chap. 9.

[7]It can be seen that the expression $\mathrm{Tr}_o(\hat{a}_\nu^\dagger\hat{a}_\nu\hat{\rho}_o^{\tau-t}(\tau))$ is equal to $N_\nu(\tau) = \mathrm{Tr}(\hat{N}_\nu\hat{\rho}(\tau))$ by using Eq. (5.12), permuting factors in the trace and noting that $[\hat{H}_0, \hat{N}_\nu] = 0$. It is also trivial to verify that $N_\nu(\tau) = \mathrm{Tr}_o(\hat{N}_\nu\hat{\rho}_o(\tau))$. After splitting the full trace in the definition $N(\tau) = \mathrm{Tr}(\hat{N}_\nu\hat{\rho}(\tau))$, we

expression, Eq. (5.21) becomes

$$\hat{\mu}_\nu(t) = \frac{i}{M_\nu \omega_\nu} \int_0^t \left( N_\nu(\tau) + \frac{1}{2} \right) [\hat{F}_\nu^{\tau-t}, \hat{\rho}_e^{\tau-t}] \cos \omega_\nu(\tau - t) \, d\tau$$
$$- \frac{1}{2 M_\nu \omega_\nu} \int_0^t \{\hat{F}_\nu^{\tau-t}, \hat{\rho}_e^{\tau-t}\} \sin \omega_\nu(\tau - t) \, d\tau \tag{5.27}$$

Correspondingly, part of the first term of $\hat{\lambda}_\nu(t)$ from Eq. (5.25) can be approximated to

$$\text{Tr}_0 \sum_{\nu'=1}^{N_o} \left( \{\hat{P}_\nu, \hat{X}_{\nu'}^{\tau-t}\} \hat{\rho}^{\tau-t}(\tau) \right) \simeq \text{Tr}_0 \left( \{\hat{P}_\nu, \hat{X}_\nu^{\tau-t}\} \hat{\rho}_0^{\tau-t}(\tau) \right) \hat{\rho}_e^{\tau-t}(\tau)$$
$$\simeq \hbar \left( 2 N_\nu(\tau) + 1 \right) \sin \omega_\nu(\tau - t) \, \hat{\rho}_e^{\tau-t}(\tau) \tag{5.28}$$

with the use of Eqs. (5.19) and (5.8). After some rearrangements, Eq. (5.25) is approximated to

$$\hat{\lambda}_\nu(t) = -\frac{1}{i} \int_0^t \left( N_\nu(\tau) + \frac{1}{2} \right) [\hat{F}_\nu^{\tau-t}, \hat{\rho}_e^{\tau-t}(\tau)] \sin \omega_\nu(\tau - t) \, d\tau$$
$$+ \frac{1}{2} \int_0^t \{\hat{F}_\nu^{\tau-t}, \hat{\rho}_e^{\tau-t}(\tau)\} \cos \omega_\nu(\tau - t) \, d\tau. \tag{5.29}$$

With these approximations we are one step closer to a system of equations of motion that can be simulated in real time. There is one last obstacle to overcome before we can include $\hat{\mu}_\nu(t)$ and $\hat{\lambda}_\nu(t)$ in $\hat{\rho}_e(t)$ and $N_\nu(t)$: how to simulate the time integrals. The integrals appearing in Eqs. (5.27) and (5.29) present strong similarities. They contain commutators and anticommutators of the same quantities and some oscillator-dependent phases. Their evolution can be coupled and controlled by a new set of operators, as we see in next section.

## 5.4 ECEID's Equations of Motion

We introduce four auxiliary electronic operators $(\hat{C}_\nu^c, \hat{A}_\nu^c, \hat{C}_\nu^s, \hat{A}_\nu^s)$ for every oscillator, defined[8] as

---

apply the electronic trace to the full system DM $N_\nu(\tau) = \text{Tr}_0 \left( \hat{N}_\nu \text{Tr}_e(\hat{\rho}(\tau)) \right)$, hence the proposition is verified.

[8]The logic behind the naming of these operators takes into account that the operators $\hat{C}_\nu^{c,s}$ contain a commutator, whereas $\hat{A}_\nu^{c,s}$ have an anticommutator; $\hat{C}_\nu^c, \hat{A}_\nu^c$ present a cosine and $\hat{C}_\nu^s, \hat{A}_\nu^s$ a sine.

$$\hat{C}_\nu^c(t) = \int_0^t \left(N_\nu(\tau) + \tfrac{1}{2}\right) [\hat{F}_\nu^{\tau-t}, \hat{\rho}_e^{\tau-t}(\tau)] \, \cos\omega_\nu(\tau - t) \, d\tau \qquad (5.30)$$

$$\hat{A}_\nu^c(t) = \frac{1}{2} \int_0^t \{\hat{F}_\nu^{\tau-t}, \hat{\rho}_e^{\tau-t}(\tau)\} \, \cos\omega_\nu(\tau - t) \, d\tau \qquad (5.31)$$

$$\hat{C}_\nu^s(t) = \int_0^t \left(N_\nu(\tau) + \tfrac{1}{2}\right) [\hat{F}_\nu^{\tau-t}, \hat{\rho}_e^{\tau-t}(\tau)] \, \sin\omega_\nu(\tau - t) \, d\tau \qquad (5.32)$$

$$\hat{A}_\nu^s(t) = \frac{1}{2} \int_0^t \{\hat{F}_\nu^{\tau-t}, \hat{\rho}_e^{\tau-t}(\tau)\} \, \sin\omega_\nu(\tau - t) \, d\tau. \qquad (5.33)$$

In terms of these operators, Eqs. (5.27) and (5.29) can be written in a very compact way as

$$\hat{\mu}_\nu(t) = \frac{1}{M_\nu \omega_\nu} (i\,\hat{C}_\nu^c(t) - \hat{A}_\nu^s(t)) \qquad (5.34)$$

$$\hat{\lambda}_\nu(t) = i\hat{C}_\nu^s(t) + \hat{A}_\nu^c(t). \qquad (5.35)$$

Their time dependency is entirely described by the auxiliary operators.

The evolution of the auxiliary operators can be obtained simply by taking the time derivative of their definition. The resulting EOM

$$\dot{\hat{C}}_\nu^c(t) = -\frac{i}{\hbar} [\hat{H}_e, \hat{C}_\nu^c(t)] + \omega_\nu \hat{C}_\nu^s(t) + (N_\nu(t) + \tfrac{1}{2})[\hat{F}_\nu, \hat{\rho}_e(t)] \qquad (5.36)$$

$$\dot{\hat{C}}_\nu^s(t) = -\frac{i}{\hbar} [\hat{H}_e, \hat{C}_\nu^s(t)] - \omega_\nu \hat{C}_\nu^c(t) \qquad (5.37)$$

$$\dot{\hat{A}}_\nu^c(t) = -\frac{i}{\hbar} [\hat{H}_e, \hat{A}_\nu^c(t)] + \omega_\nu \hat{A}_\nu^s(t) + \frac{1}{2} \{\hat{F}_\nu, \hat{\rho}_e(t)\} \qquad (5.38)$$

$$\dot{\hat{A}}_\nu^s(t) = -\frac{i}{\hbar} [\hat{H}_e, \hat{A}_\nu^s(t)] - \omega_\nu \hat{A}_\nu^c(t) \qquad (5.39)$$

are coupled and can be solved numerically. They all contain a Liouville-like commutator term and an harmonic term. The EOM of operators $\hat{C}_\nu^c$ and $\hat{A}_\nu^c$ also present driving terms that depend on the electron–phonon coupling and the electronic DM. We discuss the physical meaning of these terms in Sect. 7.3.

If the dynamics of $\hat{\mu}_\nu(t)$ and $\hat{\lambda}_\nu(t)$ is entirely determined by the auxiliary operators, the dynamics of $\hat{\rho}_e(t)$ and $N_\nu(t)$ depends on $\hat{\mu}_\nu(t)$ and $\hat{\lambda}_\nu(t)$. We rewrite their EOM here for completeness

$$\dot{\hat{\rho}}_e(t) = \frac{1}{i\hbar} [\hat{H}_e, \hat{\rho}_e(t)] - \frac{1}{i\hbar} \sum_{\nu=1}^{N_o} [\hat{F}_\nu, \hat{\mu}_\nu(t)] \qquad (5.40)$$

$$\dot{N}_\nu(t) = \frac{1}{\hbar M_\nu \omega_\nu} \mathrm{Tr}_e\left(\hat{F}_\nu \hat{\lambda}_\nu(t)\right). \qquad (5.41)$$

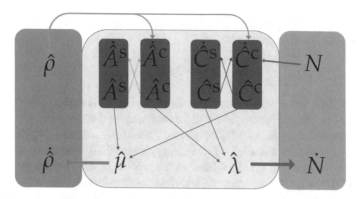

**Fig. 5.1** Schematic representation of the EOM and their mutual dependency

The set of Eqs. (5.34)–(5.41) represents the required closed system of equations which, given an initial condition, can be integrated in time to simulate the evolution of the whole system. In Fig. 5.1 we sketch the linked time evolution of the quantities in ECEID, to give an idea of how ECEID dynamics unfolds. These EOM conserve total energy, as we prove in Sect. (5.6).

## 5.5 From Many-Electron to One-Electron Equations of Motion

Throughout the derivation above we did not make any assumption about the form of the electronic operators that can be as general as one requires and can, in principle, involve many-body operators.

To be able to apply the method to systems with large numbers of DOF and simulate them, we must express the EOM in one-electron form. We do this by tracing out all but one electron, with the application of $N_e \text{Tr}_{e,2,\dots,N_e}$, where $N_e$ is the total number of electrons in the system. We choose $\hat{H}_e$ and $\hat{F}_\nu$ to be one-body operators, neglecting electron-electron interaction. Because of this, all other electronic operators in the EOM can be replaced by their one-electron counterparts, except for the anticommutator term in Eq. (5.38) $\{\hat{F}_\nu, \hat{\rho}_e\}$.

Following [1], that term transforms into

$$\{\hat{F}_\nu^{(1)}(1), \hat{\rho}_e^{(1)}(1)\} + 2\text{Tr}_{e,2}\left(\hat{F}_\nu^{(1)}(2)\hat{\rho}_e^{(2)}(1,2)\right) \tag{5.42}$$

where superscripts $^{(1)}$ and $^{(2)}$ denote respectively one- and two-electron operators. The simplest decoupling for the two-particle DM is

$$\hat{\rho}_e^{(2)}(12, 1'2') = \hat{\rho}_e^{(1)}(11')\hat{\rho}_e^{(1)}(22') - \hat{\rho}_e^{(1)}(12')\hat{\rho}_e^{(1)}(21'), \tag{5.43}$$

which is valid for independent electrons. Using this decomposition in Eq. (5.42), we obtain the one-electron expression

$$\{\hat{F}_v, \hat{\rho}_e(t)\} - 2\hat{\rho}_e(t)\hat{F}_v\hat{\rho}_e(t), \tag{5.44}$$

where now $\hat{\rho}_e(t)$ is the one-electron DM. In Appendix A the derivation of the many-electron to one-electron projection is discussed in more detail.

In the derivation of Eq. (5.44), we ignored the additional term $\hat{\rho}_e(t) \, \text{Tr}_e(\hat{F}_v\hat{\rho}_e(t))$ that corresponds to the so-called "Hartree" diagram in NEGF treatments of electron–phonon interactions [3], and is related to the motion of the oscillator centroid, a mean-field property. The accuracy of (5.43) reduces with increasing electron–phonon coupling; further corrections to this approximation are discussed in [4]. Screening can be included in a one-electron mean-field picture within a Hartree-Fock scheme following [4], or in a time-dependent density-functional framework [5].

## 5.6  Total Energy Conservation

We introduce the total energy $E = E_e + E_o + E_c$, where $E_e = \text{Tr}_e(\hat{H}_e\hat{\rho}_e(t))$, $E_o = \sum_v \hbar\omega_v(N_v(t) + 1/2)$ and $E_c = -\sum_v \text{Tr}_e(\hat{F}_v\hat{\mu}_v(t))$, and we show that it is identically conserved by ECEID's EOM.

The time-derivative of the total energy of the system is

$$\dot{E} = \text{Tr}_e(\hat{H}_e\dot{\hat{\rho}}_e(t)) + \sum_{v=1}^{N_o} \left(\hbar\omega_v\dot{N}_v(t) - \text{Tr}_e(\hat{F}_v\dot{\hat{\mu}}_v(t))\right). \tag{5.45}$$

Plugging Eq. (5.5) into the first term of Eq. (5.45) and using Eq. (5.34), we get

$$-\frac{1}{M_v\hbar\omega_v}\text{Tr}_e\left([\hat{F}_v, \hat{C}_v^c]\hat{H}_e + i[\hat{F}_v, \hat{A}_v^s]\hat{H}_e\right). \tag{5.46}$$

With Eq. (5.41), the second term of Eq. (5.45) becomes

$$\frac{1}{M_v}\text{Tr}_e\left(i\hat{F}_v\hat{C}_v^s + \hat{F}_v\hat{A}_v^c\right). \tag{5.47}$$

Using the time derivative of Eq. (5.34) together with Eqs. (5.36) and (5.39), the third term of Eq. (5.45) can be written as

$$-\frac{1}{M_v\hbar\omega_v}\text{Tr}_e\left(\hat{F}_v\left[\hat{H}_e, \hat{C}_v^c\right] + i\hat{F}_v\left[\hat{H}_e, \hat{A}_v^s\right]\right) - \frac{1}{M_v}\text{Tr}_e\left(i\hat{F}_v\hat{C}_v^s + \hat{F}_v\hat{A}_v^c\right). \tag{5.48}$$

Summing (5.46), (5.47) and (5.48) we obtain total energy conservation $\dot{E} = 0$.

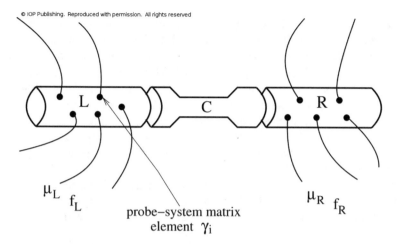

**Fig. 5.2** A typical system where the OB method is applied [6]: a central region of a nanowire is coupled to its left and its right to leads that, in turn, are connected to probes with a set chemical potentials $\mu_{L/R}$ and population distributions $f_{L/R}$

## 5.7   Open Boundaries in ECEID

In the derivation above, the number of electrons is fixed. The ECEID's EOM describe the evolution of a closed system. It is possible to allow electron injection and extraction by coupling the system to external reservoirs. We implement an Open Boundaries (OB) setup by following the multiple probes OB method derived in [6]. More details about its derivation and its implementation can be found in Appendix B.

We consider a nanowire, whose central region is connected to a left and a right lead, as shown in Fig. 5.2. The central region contains the vibrational DOF and is the area where the dynamical scattering and the electron–phonon energy exchanges occur. The leads are usually metallic and some of their sites are connected to external probes, acting as particle baths. This setup effectively screens the finite size of the leads by broadening their discrete levels into a continuous spectrum. In the wide band limit for the external baths [6], the system's embedding self energy is $\hat{\Sigma}^{\pm} = \mp(\mathrm{i}\Gamma/2)(\hat{I}_L + \hat{I}_R)$ where $\Gamma$ sets the coupling to the baths and $\hat{I}_{L/R}$ are the identity operators over the left/right regions coupled to the baths. We are using an orthonormal real space basis throughout.

The introduction of the OB transforms Eq. (5.5) into [6]

$$i\hbar\dot{\hat{\rho}}_e(t) = [\hat{H}_e, \hat{\rho}_e(t)] - \sum_{\nu=1}^{N_o}[\hat{F}_\nu, \hat{\mu}_\nu(t)] + \underbrace{\hat{\Sigma}^+\hat{\rho}_e(t) - \hat{\rho}_e(t)\hat{\Sigma}^-}_{\text{extraction}}$$

$$+ \underbrace{\int_{-\infty}^{\infty}\Big(\hat{\Sigma}^<(E)\hat{G}_S^-(E) - \hat{G}_S^+(E)\hat{\Sigma}^<(E)\Big)dE}_{\text{injection}} \qquad (5.49)$$

where $\hat{\Sigma}^+ \hat{\rho}_e(t) - \hat{\rho}_e(t)\hat{\Sigma}^-$ represents electron extraction and the last term portrays electron injection. The injection integral involves the retarded (advanced) Green's function for the lead-sample-lead system $\hat{G}_S^{+(-)}(E)$ in the presence of the baths and the quantity $\hat{\Sigma}^<(E) = \frac{\Gamma}{2\pi}(f_L(E)\hat{I}_L + f_R(E)\hat{I}_R)$, where $f_{L(R)}(E)$ are the desired incoming electronic distributions. These could describe a conventional applied electrochemical bias or also electron beams targeted at specific energies.

In Appendix B, we show an explicit form of the injection integral for zero temperature electronic distributions. In the examples in the following chapters we choose $f_{L(R)}(E)$ to be fixed at zero temperature, but it would be possible to choose them at a finite temperature. If $f_L(E)$ and $f_R(E)$ had different temperatures, ECEID could be used to simulate electron injection due to a thermal imbalance between the reservoirs. At a later stage, in Sect. 6.2, we will introduce a damping mechanism of the auxiliary operators based on $\Gamma$, to mimick extended systems without the extra cost.

## 5.8   Implementing ECEID in a Computer Simulation

The ECEID method has been implemented in a Fortran 90 code called *ElPh* that is available in the repository https://bitbucket.org/gilgalad/eceid. It integrates the ECEID EOM using a leapfrog algorithm, which, for stability purposes, is alternated with a single Euler integration step (typically every 100 timesteps). The timestep is imposed by the very fast electronic dynamics and typically is of the order $\simeq 1$ attosecond, whereas our simulations intend to probe the time evolution of phonons, which can require times exceeding 1 picosecond.

The need to simulate millions of timesteps makes the code's efficiency a priority, therefore, optimizing the expensive operations within each timestep is an essential requirement. The most time consuming operation in ECEID is matrix-matrix multiplication, which normally has a $\mathcal{O}(n^3)$ cost for dense $n \times n$ matrices. The electron–phonon coupling operator $\hat{F}$ appears often in the EOM and, in our calculations, it usually features only a few non-zero elements, in the real-space basis. To speed up matrix-matrix multiplications, we developed ad hoc routines that involve sparse matrices, such as $\hat{F}$ or $\hat{H}_e$. We store the non-zero elements of the sparse matrices and their coordinates and we compute multiplications by determining only the non-zero components of the final matrix. These routines tend to scale as $\mathcal{O}(n^2)$ and greatly reduce the code's computational time. The speeding up effect when compared to standard multiplication routines is substantial, especially for large matrices.

One of the motivations that has driven the development of ECEID is to have a method that scales well with the number of oscillators. Therefore it is paramount for *ElPh* to deal with many oscillators in an efficient way. Each oscillator is defined in a structure, a feature of Fortran analogous to an object in C++. Thanks to the form of ECEID's EOM, OpenMP could be employed to make the code parallel. During every timestep, the dynamics of each oscillator-specific variable is computed independently in a thread. When all the oscillator variables are determined, the electronic

DM is evolved and the program goes one timestep forward. The linear scaling with the number of oscillators is an intrinsic feature of the method. This procedure can produce some overhead if an oscillator's variables take more time than the other ones to compute. We investigate the impact of overhead and the code's performance in Sect. 6.5.

The shared memory design of the OpenMP instructions makes the code usable only on single machines. A thread has access to a global memory area where the global variables are stored, such as constants or the electronic DM $\hat{\rho}_e(t)$, and a local memory where there are oscillator $\nu$ specific quantities. The threads have different read and write restrictions on the memory, but in OpenMP both the global and the local memory lie in partitions of the memory of the same machine. This shared memory design is efficient when there is a frequent data transfer between threads, such as at the end of every timestep, when the evolution of $\hat{\rho}_e(t)$ is determined by the dynamics of all the oscillator-specific $\hat{\mu}_\nu(t)$.

Another widely employed solution to parallelism is a private memory design such as MPI. Codes with MPI can be run by different processes on different machines and every process has a private amount of memory to its exclusive use. Processes communicate with each other through data transfer that is achieved by specific instructions. In principle, *ElPh* can be implemented in MPI, but it would be very challenging. The frequent data exchange at the end of every short timestep could easily become a bottleneck and would require a very careful optimization of the communication between processes. A further exciting possibility to implement parallelism efficiently is the use of coarrays in Fortran 2003.

A performance analysis of the code is provided in Sect. 6.5, with a focus on the scaling with a varying number of electronic sites and oscillators.

### 5.8.1 Code Breakdown

Here follows a schematic breakdown of the program components and their functionality. A typical simulation consists of compiling the program with *make*, executing the geometry setup with *./diagH* and running the program with *./elph*. I invite those who are interested in the method and its implementation to contact me at vrizzi01@qub.ac.uk.

**modvar.f90** contains the system's parameters that determine its geometry and details of the observables that can be measured during the simulation. For example, there is the number of electronic sites and oscillators, and the sites where the current is evaluated. It also stores physical constants and structure definitions. Changing this file requires the program to be recompiled with *make*.

**input.dat** keeps data involving different oscillators on different lines. The 4 columns of line $\nu$ in the file correspond respectively to oscillator $\nu$'s electron–phonon coupling modulus $F_\nu$ (in eV/Å), initial oscillator occupation $N_\nu(0)$, frequency $\hbar\omega_\nu$ (in eV) and mass $M_\nu$ (in a.m.u.). This data is read at the start of every execution of the program. A change in this file does not require a recompilation of the program.

**diagH.f90**    this program, when executed, sets up the geometry of the system and stores it in files that will be read by the main program elph.f90. It creates matrices with $\hat{H}_e$, $\hat{F}_v$ and $\hat{\rho}_e(0)$. It must be always executed before elph.f90.

**elph.f90**    is the main program where the ECEID EOM are evolved in time and the observables are evaluated. For example, in a thermalization simulation the electronic and oscillator temperatures are measured, while in an electron injection simulation the observables typically are the current, the mean oscillator occupation and the electronic levels occupation.

**steady.f90**    here some accessory quantities are evaluated such as the system's elastic transmission, the elastic current, and the local density of states.

**constx3l.f90**    contains the parameters for the exact calculations proposed in Chap. 6. There ECEID is applied to problems containing a few degrees of freedom and compared to exact solutions.

**x3l.f90**    is the program where the exact dynamics of small electron–phonon systems is solved in the presence of 1 oscillator, with a truncation in $N$ space.

**x3l2o.f90**    solves the electron–phonon problem exactly for geometries with 2 oscillators.

**toydiffusion.f90**    contains a simulation of the kinetic model presented in Chap. 7. It is used to rationalize the dynamics of the electronic levels during thermalization.

**makefile**    compiles the set of programs.

**Releasenotes.txt**    keeps track of the changes in the different releases of the code and briefly explains how to use the new features.

# References

1. Horsfield, A.P., D.R. Bowler, A.J. Fisher, T.N. Todorov, and C.G. Sánchez. 2004. Beyond Ehrenfest: Correlated non-adiabatic molecular dynamics. *Journal of Physics: Condensed Matter* 16 (46): 8251–8266. https://doi.org/10.1088/0953-8984/16/46/012.
2. Horsfield, A.P., D.R. Bowler, H. Ness, C.G. Sánchez, T.N. Todorov, and A.J. Fisher. 2006. The transfer of energy between electrons and ions in solids. *Reports on Progress in Physics* 69 (4): 1195–1234. https://doi.org/10.1088/0034-4885/69/4/R05.
3. Frederiksen, T., M. Paulsson, M. Brandbyge, and A.-P. Jauho. 2007. Inelastic transport theory from first principles: Methodology and application to nanoscale devices. *Physical Review B* 75 (20): 205413. https://doi.org/10.1103/PhysRevB.75.205413
4. Horsfield, A.P., D.R. Bowler, A.J. Fisher, T.N. Todorov, and C.G. Sanchez. 2005. Correlated electron-ion dynamics: The excitation of atomic motion by energetic electrons. *Journal of Physics: Condensed Matter* 17 (30): 4793–4812. https://doi.org/10.1088/0953-8984/17/30/006.
5. Burke, K., R. Car, and R. Gebauer. 2005. Density functional theory of the electrical conductivity of molecular devices. *Physical Review Letters* 94 (14): 146803. https://doi.org/10.1103/PhysRevLett.94.146803
6. McEniry, E.J., D.R. Bowler, D. Dundas, A.P. Horsfield, C.G. Sánchez, and T.N. Todorov. 2007. Dynamical simulation of inelastic quantum transport. *Journal of Physics: Condensed Matter* 19 (19): 196201. https://doi.org/10.1088/0953-8984/19/19/196201

# Chapter 6
# ECEID Validation

In this chapter we validate ECEID by examining a number of applications to test it. First, we employ an exact simulation and compare results with ECEID on small closed systems to check the range of validity of the approximations in the method and to find limits where ECEID approaches the exact solution. Then we validate the open boundaries setup by recovering the Landauer two terminal limit for a perfect nanowire and by testing Joule heating. By varying the wire length and keeping a constant density of oscillators, we recover a Ohm's law microscopically and compare it to perturbative results. We add elastic scattering to the picture with the inclusion of onsite disorder. At last, we test ECEID's scaling with performance tests.

## 6.1 Comparison with an Exact Simulation

The size of the one-electron full quantum problems described by the model Hamiltonian (5.1) is determined by the number of electronic states $N_e$ and the number of oscillators $N_o$. After setting a cutoff $N_C$ in phonon space,[1] the finite size of the system operators is $s = N_e \cdot N_C^{N_o}$. For systems with a few degrees of freedom, $s$ is not too large and exact simulations can be employed to evaluate the dynamics of the system. Obviously, such problems can also be simulated with ECEID. The possibility of comparing exact simulations with ECEID allows one to explore the ECEID approximations and the range of their validity.

The exact evolution of the initial full DM $\hat{\rho}(0)$ is determined by

$$\hat{\rho}_{ex}(t) = e^{-\frac{i}{\hbar}\hat{H}t}\hat{\rho}(0)e^{\frac{i}{\hbar}\hat{H}t} \tag{6.1}$$

---

[1] $N_C$ is the truncation of the oscillator's basis and represents the maximum allowed $N_\nu(t)$, for any $\nu$.

© Springer International Publishing AG, part of Springer Nature 2018
V. Rizzi, *Real-Time Quantum Dynamics of Electron-Phonon Systems*,
Springer Theses, https://doi.org/10.1007/978-3-319-96280-1_6

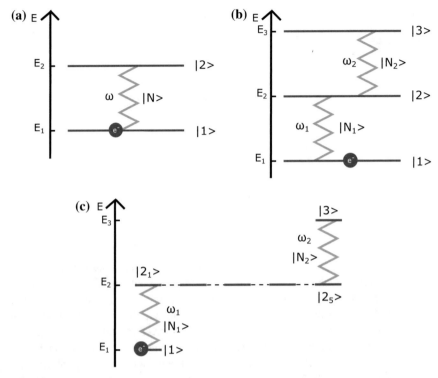

**Fig. 6.1** Sketches of the systems used in the comparison tests between ECEID and an exact simulation. In **a**, a 2-level system coupled to a single oscillator. In **b**, a 3-level system with two oscillators coupling level 1 to level 2 and level 2 to level 3. In **c**, a multi-level system that takes the system in panel (**b**) and transforms its level 2 into a chain of 5 coupled levels. In all cases, the initial condition on the electron is on level 1

for all times $t$. The numerical simulation of (6.1) requires the diagonalization of $\hat{H}$ and the projection of the initial $\hat{\rho}(0)$ on the Hamiltonian's eigenstates. The quality of the solution depends on the truncation in phonon space, therefore convergence tests in $N_C$ are necessary to reach the desired level of precision. For large systems, the computational cost of this approach soon becomes unmanageable as the computational size scales exponentially with the number of oscillators. As a rule of thumb, a tractable choice is $s \simeq 1000$. In Fig. 6.1, we show a sketch of the systems that we employ in the examples below.[2] They include at most 2 oscillators.

The limitation in system size is not a problem for the purpose of validating ECEID. Small quantum systems present high levels of coherence that are very challenging to reproduce for approximate methods. ECEID's validation focuses on the ability of the method to capture the dynamics of inelastic electronic transitions. The small

---

[2]The initial condition mentioned in Fig. 6.1 limits the initial response to just one process, phonon absorption, and was chosen for consistency among the different systems tested. Starting from an upper level would allow an initial phonon emission.

systems in Fig. 6.1 represent an ideal test bed. In the examples, the oscillators always have $M = 0.5$ amu and $\hbar\omega = 0.2$ eV.

### 6.1.1 An Exact Limit on a 2-Level System with 1 Oscillator

The simplest system in the test is shown in Fig. 6.1a and is a 2-level system. Two electronic levels $|1\rangle$ and $|2\rangle$ with respective onsite energies $E_1 = 0$ and $E_2 = E$ are coupled to an oscillator by $\hat{F} = F(|1\rangle\langle 2| + |2\rangle\langle 1|)$. The initial phonon mode occupation is $N(0) = N$.

When the levels and the oscillator are in resonance $E = \hbar\omega = 0.2$ eV and in the limit $F \longrightarrow 0$, the eigenspectrum of the full system consists of degenerate couples of states $|a\rangle = |1, N\rangle$ and $|b\rangle = |2, N - 1\rangle$. When $F$ is finite, the electron phonon coupling $\hat{V} = -\hat{F}\hat{X}$ generates a coupling between the levels

$$V_{ab} = \langle a|\hat{V}|b\rangle = -F\sqrt{\frac{\hbar N}{2M\omega}} \qquad (6.2)$$

that breaks the degeneracy. Provided that $F$ is small, the electron phonon coupling acts as a small perturbation $V_{ab} \ll E$ and the most relevant states in the dynamics of the system still consist of states $|a\rangle$ and $|b\rangle$. Such systems can be solved analytically and their solution oscillates harmonically between the 2 states with a Rabi frequency $\omega_R = |V_{ab}|/\hbar$.

We performed simulation starting from state $|a\rangle$ for a number of $F$ and $N$. For example, in Fig. 6.2a is shown a resonant case with $F = 0.05$ eV/Å and $N = 1$. The exact simulation displays complete oscillations of population at a frequency $\omega_R$, while ECEID shows long lived oscillations with a slightly faster frequency and less complete population oscillations, when compared to the exact case.

It is natural to wonder if there is a limit in which ECEID tends to the exact solution. Assuming that $|V_{ab}| \ll E$, the system's dynamics can be described by the following time-dependent linear combination $a(t)|a\rangle + b(t)|b\rangle$. Checking the exact form of $\hat{\mu}(t)$ in Eq. (5.21), it is possible to see that, if the full DM contains only oscillator states $|N\rangle$ and $|N - 1\rangle$ (as is the case for $|a\rangle$ and $|b\rangle$), the operators $\hat{a}\hat{a}$ and $\hat{a}^\dagger\hat{a}^\dagger$ that arise from $\{\hat{X}, \hat{X}^{\tau-t}\}$ give zero contribution.

Therefore, after the application of the oscillator trace, the first term of Eq. (5.21) can be written exactly as

$$\frac{i}{M\omega} \int_0^t e^{\frac{i}{\hbar}\hat{H}_e(\tau-t)} \left[\hat{F}, \left(N(\tau) + \frac{1}{2}\right)|a(\tau)|^2|1\rangle\langle 1| + \left(N(\tau) - \frac{1}{2}\right)|b(\tau)|^2|2\rangle\langle 2|\right]$$
$$e^{-\frac{i}{\hbar}\hat{H}_e(\tau-t)} \cos(\omega(\tau - t))\, d\tau. \qquad (6.3)$$

By comparing Eq. (6.3) with its ECEID equivalent (the first term of Eq. (5.27)), we see that the two are identical in the limit of large $N$. A similar reasoning can be

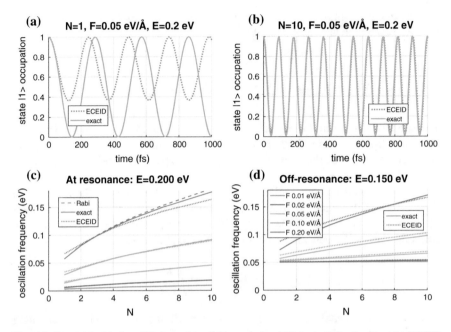

**Fig. 6.2** Panel (**a**)–(**b**) show the dynamics of electronic level $|1\rangle$ in the 2-level system for ECEID (dotted blue) and the exact solution (solid green). In panel (**a**) the phonon starts from $N = 1$, while in panel (**b**) from $N = 10$. Panel (**c**) displays the population oscillation main frequency for a range of $F$ and $N$ in ECEID (dotted) and in exact simulations (solid) of a resonant system. They are compared with the relative Rabi frequencies (dashed). Panel (**d**) is the same as (**c**), for simulations of an off-resonant system with $E = 0.150\,\text{eV}$

applied to the EOM for $\dot{N}$. The limit of small $V_{ab}$ and large $N$ is a clear strategy for systematic convergence to the solution case on a 2-level system. In Fig. 6.2b we test the same system as in (a), with $N = 10$ and we see that ECEID dynamics is indeed converged and superimposable to the exact case.

Next, we test several combinations of $F$ and $N$, derive the main oscillation frequency with a Fourier transform of the level dynamics, and in Fig. 6.2c show a comparison between ECEID, exact simulations and Rabi frequencies. The level coupling is proportional to $F$ and $\sqrt{N}$, so, for the limit to be satisfied, a balance must be reached between $F$ and $N$. For each $F$, there is a best $N$ that strikes a compromise between a low enough $V_{ab}$ and a high enough $N$.

For low $F$ ($F = 0.01\,\text{eV}/\text{Å}$ and $F = 0.02\,\text{eV}/\text{Å}$) the frequency agreement is good throughout the $N$ range. For $F = 0.05\,\text{eV}/\text{Å}$, the optimal $N$ is about $N = 10$, as Fig. 6.2b indicated. For higher $F$ ($F = 0.10\,\text{eV}/\text{Å}$ and $F = 0.20\,\text{eV}/\text{Å}$) the optimal $N$ decreases, becoming respectively about $N = 5$ and $N = 3$. We notice that in most cases the Rabi frequency matches the exact frequency, except for the case of high $F$ and $N$. In that case $V_{ab}$ is so large that the two-level assumption breaks down

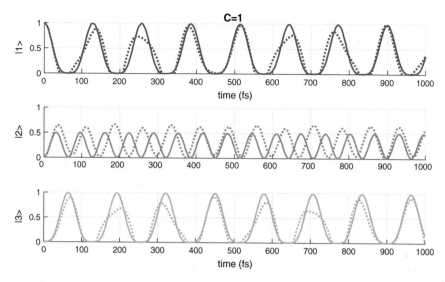

**Fig. 6.3** ECEID (dotted) and exact (solid) simulations of a 3-level system with 2 oscillators as sketched in Fig. 6.1b. $F_{1,2} = 0.05$ eV/Å and $N_{1,2} = 10$. The dynamics of electronic level $|1\rangle$ is in the top panel, level $|2\rangle$ in the central one and level $|3\rangle$ in bottom one

and more levels become relevant to the dynamics. The system dynamics cannot be described by 2-level Rabi oscillations any more.

We perform the same set of simulations for an off-resonance case with $E = 0.150$ eV and in Fig. 6.2d show the main oscillation frequencies from ECEID and the exact simulations. For low $F$ the dominant transition mechanism has a frequency close to the off-resonance offset $\hbar\omega - E = 0.05$ eV for all considered $N$. For a higher $F$, the frequencies increase and there is a shift between the main frequency in ECEID and the exact simulations, with the frequency in ECEID being higher. For $F = 0.20$ eV/Å the electron-phonon coupling becomes again the dominant mechanism and the frequencies from both simulations show a behaviour comparable to the resonant case.

## 6.1.2 Extension to a 3/Many-Level System with 2 Oscillators

The 2-level system offered the possibility to test two of the three ECEID approximations: the decoupling of the DM and the omission of double (de)excitations. To include the remaining approximation in the picture, i.e. the omission of explicit interaction between different oscillators (the second approximation in Sect. 5.3), we need to introduce at least one other oscillator in the system.

The simplest realization is a 3-level system with energies $E_1 = -E = -0.2$ eV, $E_2 = 0$, $E_3 = E = 0.2$ eV and two oscillators coupling level 1 to level 2 and level

2 to level 3, as drawn in Fig. 6.1b. The oscillators are identical and have electron-phonon coupling operators $\hat{F}_1 = F_1(|1\rangle\langle 2| + |2\rangle\langle 1|)$ and $\hat{F}_2 = F_2(|2\rangle\langle 3| + |3\rangle\langle 2|)$, in analogy with the previous case.

We simulate a scenario equivalent to the one where there was the best agreement in the previous case: $F_{1,2} = 0.05$ eV/Å and initial occupations $N_{1,2} = 10$. In Fig. 6.3 we show the dynamics of the electronic levels. States $|1\rangle$ and $|3\rangle$ display a good agreement between ECEID and the exact simulation in terms of oscillation frequency and, to a lesser degree, of population occupation. The agreement for state $|2\rangle$ is not as close.

Similarly to the 2-level case, it is possible to solve the system analytically to determine a limit where ECEID converges to the exact case. Assuming again that the electron-phonon coupling energy scale is smaller than the level spacing $F_{1,2}\sqrt{\frac{\hbar N_{1,2}}{2M\omega}} \ll E$, the system eigenstates condense into groups made of 3 levels $|a\rangle = |1, N_1, N_2\rangle$, $|b\rangle = |2, N_1 - 1, N_2\rangle$, $|c\rangle = |3, N_1 - 1, N_2 - 1\rangle$. An analytic solution for such a system is a linear combination of these states $a(t)|a\rangle + b(t)|b\rangle + c(t)|c\rangle$. Checking the exact Eq. (5.21), it is evident that mixed operators $\hat{a}_1\hat{a}_2$ and $\hat{a}_1^\dagger\hat{a}_2^\dagger$ appear and have a non-zero contribution.

Proceeding with the derivation and comparing the ECEID form of $\hat{\mu}_{1/2}$ with the exact one, we notice that the previous term (6.3) appears again in the exact case, together with other terms[3] proportional to $a^*(t)c(t)$ or $a(t)c^*(t)$ that are not present in ECEID. To make ECEID converge to the exact case, all extra terms must go to zero. The previous condition of low $F$ and high $N$ is still valid, but it is not enough to have exact convergence here, as the results in Fig. 6.3 highlighted.

The easiest way to reduce the influence of the extra terms, is to devise system geometries where the simultaneous occupation of state $|a\rangle$ and $|c\rangle$ is minimal. Starting from the 3-level system under investigation here, a possibility is to transform electronic level 2 into an N-level chain with hopping 1 eV. The chain starts on level $2_1$ and ends on site $2_C$. An electron on level 1 would inelastically hop on $2_1$ and propagate along the chain. If the chain is long enough, by the time the electron reaches the end and inelastically hops on level 3, the population of 1 would be largely depleted.

---

[3] The extra terms for $\hat{\mu}_1$ are

$$\frac{i}{2M\omega}F_1 \int_0^t d\tau (N(\tau) + 1)\cos(\omega(\tau - t))\left(c^*(\tau)a(\tau)e^{i\omega(\tau - t)}|2\rangle\langle 1| - c(\tau)a^*(\tau)e^{-i\omega(\tau - t)}|1\rangle\langle 2|\right)$$

$$+\frac{1}{2M\omega}F_1 \int_0^t d\tau (N(\tau) + 1)\sin(\omega(\tau - t))\left(c^*(\tau)a(\tau)e^{i\omega(\tau - t)}|2\rangle\langle 1| + c(\tau)a^*(\tau)e^{-i\omega(\tau - t)}|1\rangle\langle 2|\right)$$

and for $\hat{\mu}_2$

$$\frac{i}{2M\omega}F_2 \int_0^t d\tau (N(\tau) + 1)\cos(\omega(\tau - t))\left(-c^*(\tau)a(\tau)e^{i\omega(\tau - t)}|3\rangle\langle 2| + c(\tau)a^*(\tau)e^{-i\omega(\tau - t)}|2\rangle\langle 3|\right)$$

$$-\frac{1}{2M\omega}F_2 \int_0^t d\tau (N(\tau) + 1)\sin(\omega(\tau - t))\left(c^*(\tau)a(\tau)e^{i\omega(\tau - t)}|3\rangle\langle 2| + c(\tau)a^*(\tau)e^{-i\omega(\tau - t)}|2\rangle\langle 3|\right).$$

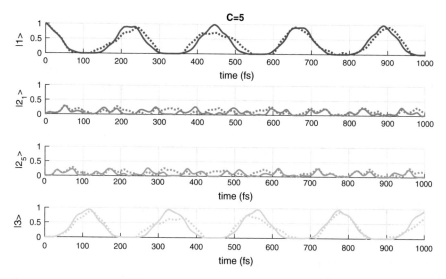

**Fig. 6.4** Comparison of ECEID (dotted) and exact (solid) simulations on a many-level system, where level 2 from the 3-level setup has been replaced by a chain of 5 levels (see Fig. 6.1c). The dynamics of level 1 is in the top panel, followed by the first level of the chain $2_1$, the last one $2_5$ and level 3 in the bottom panel

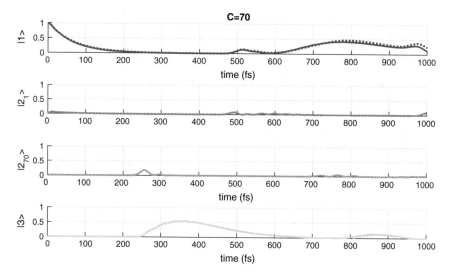

**Fig. 6.5** The same simulations as in Fig. 6.4, with $C = 70$ levels

We test this idea on a chain with $C = 5$, as sketched in Fig. 6.1, and show the resulting level dynamics in Fig. 6.4. The improvement is not very large when compared to the 3-level system in Fig. 6.3: the chain is too short. We then try a longer chain with $C = 70$ and plot the level occupations in Fig. 6.5.

In this simulation the level dynamics behaves as expected. The electron slowly leaks away from level 1. After about 250 fs, it reaches the far end of the chain and slowly populates level 3, while level 1's population is close to zero. After that, the electron is reflected back and at about 500 fs returns to level 1. The agreement between ECEID and the exact simulation is evident and very remarkable: the curves representing the level dynamics are at times so close that they cannot be distinguished. Moreover, the agreement is not limited to initial times but it carries over for long times, after many reflections and inelastic events. These simulations validate the limit derived above and provide a starting point for further simulations.

## 6.2  Mimicking an Extended System and Energy Conservation

The small systems that we described in the previous section are ideal for comparison with exact quantum solutions and could be replicated experimentally within well controlled and isolated setups, such as quantum dots [1]. The coherence of those systems is extremely sensitive to the environment: even a small external perturbation can drastically change their dynamics. ECEID has been developed to simulate extended systems that are not thought to be isolated from the environment, but interact with it. Small and controlled systems are an invaluable test bed for the inner workings of ECEID, but do not represent a typical system upon which ECEID is applied.

A typical setup where we apply ECEID is a nanowire, such as the one in Fig. 6.9 or Fig. 7.1, where a central region with phonons is coupled to leads. The leads act as an environment where the central region is immersed. In principle they should be semi-infinite, but, for computational reasons, they must be finite. The longer they are, the finer their energy level spacing and the better they resemble semi-infinite systems. To help mimic an extended (infinitely large) system, without the extra cost from extremely long leads, we introduce a quantity that takes care of the decoherence introduced by the environment in the leads and provides the needed level-broadening.

We replace $[\hat{H}_e, \hat{Q}]$ in Eqs. (5.36–5.39) by $\hat{H}_\Gamma \hat{Q} - \hat{Q} \hat{H}_\Gamma^\dagger$ where $\hat{Q} = (\hat{C}_\nu^c, \hat{A}_\nu^c, \hat{C}_\nu^s, \hat{A}_\nu^s)$, $\hat{H}_\Gamma = \hat{H}_e - i(\Gamma/2) \hat{I}'_{\text{leads}}$, $\hat{I}'_{\text{leads}}$ is the identity operator acting on (part of) the leads and $\Gamma$ a small real positive quantity. The factor $\Gamma$ is analogous to the coupling to the baths introduced in Sect. 5.7 in the OB. Both factors physically describe the same embedding of a finite system into an environment, therefore, for simplicity and physical consistency, we tend to use the same $\Gamma$ in both places. We will include $\Gamma$ in the evolution of the auxiliary operators in the examples in this Chapter and the following ones, unless otherwise stated. In the thermalization simulations in Chap. 7, we verify that once $\Gamma$ exceeds the energy-level spacing in the system, the transition rates resulting from ECEID dynamics become independent of $\Gamma$.

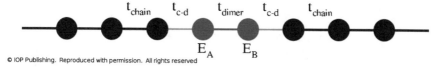

**Fig. 6.6** The test system used in [2], a resonant dimer in an atomic chain. The parameters are in the text

One could wonder if the introduction of $\Gamma$ affects the total energy conservation proved in Sect. 5.6. The total energy is in fact conserved, provided the damping self-energy and the electron-phonon coupling $\hat{F}$ lie in different subspaces. The only change in $\dot{E} = 0$ from (5.45) is the third term that, with the introduction of $\Gamma$, becomes

$$-\frac{1}{M_\nu \hbar \omega_\nu} \mathrm{Tr}_e \left( \hat{F}_\nu (\hat{H}_\Gamma \hat{C}_\nu^c - \hat{C}_\nu^c \hat{H}_\Gamma^\dagger) + i \hat{F}_\nu (\hat{H}_\Gamma \hat{A}_\nu^s - \hat{A}_\nu^s \hat{H}_\Gamma^\dagger) \right) - \frac{1}{M_\nu} \mathrm{Tr}_e \left( i \hat{F}_\nu \hat{C}_\nu^s + \hat{F}_\nu \hat{A}_\nu^c \right).$$
(6.4)

As long as $\hat{I}'_{\text{leads}} \hat{F}_\nu = 0$, $\hat{H}_\Gamma$ in Eq. (6.4) can be replaced with $\hat{H}_e$, hence total energy conservation is verified.

## 6.3 Validating the Open Boundaries

To validate the implementation of the OB setup in ECEID, we aim to reproduce results from the original paper where the OB formalism was developed [2] and to compare with the exact two-terminal Landauer solution. Such tests are purely elastic and consider a nanowire with a resonant dimer and no phonons, as in Fig. 6.6. The inelastic effects that ECEID was developed to describe are often superimposed to purely elastic effects, therefore the verification of a clear elastic limit is an essential step in the development of ECEID.

The system under analysis has leads with $N_L$ sites coupled to a central region made of $N_D = 20$ sites and a dimer on sites 10 and 11. All hoppings are $t_{\text{chain}} = -3.88$ eV, except for weak hoppings $t_{\text{c-d}} = -0.5$ eV between sites $9 - -10$ and $11 - -12$, which partially isolate the dimer from the rest of the atomic chain. The in-dimer $10 - -11$ hopping is $t_{\text{dimer}} = -3.88$ eV. All the leads' sites are coupled to the probes by $\Gamma$. A variable bias is applied to the probes, so that an IV curve can be produced as in Fig. 6.7 [2]. In the limit of long leads and small $\Gamma$, the OB curves get closer and closer to the exact Landauer case.

In analogy with Fig. 6.7, ECEID is coupled to external baths through the OB setup, to simulate Eq. (5.49) on the resonant dimer geometry. ECEID is a time-dependent simulation, so its EOM are propagated until a steady state is reached. Here, the full system's Green's Function is

$$\hat{G}_S^\pm(E) = \left( E \hat{I}_S - \hat{H}_S - \hat{\Sigma}^\pm(E) \pm i \hat{I}_S \Delta \right)^{-1}.$$
(6.5)

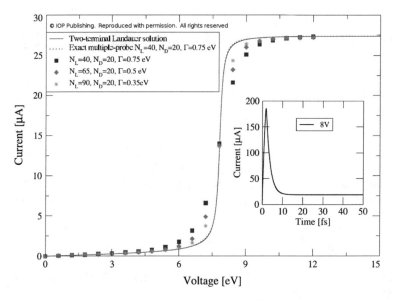

**Fig. 6.7** IV curve of the resonant dimer system from [2]. Different combinations of $N_L$ and $\Gamma$ (coloured dots) are compared to the exact Landauer result for semi-infinite leads (solid black line) and finite leads (dotted black line). In the inset, a typical current time evolution is shown with a steady state forming after about 10 fs

that, when compared to Eq. (B3) from the OB derivation, includes an extra dephasing factor $\Delta = \Gamma/2$ on all the system, in agreement with the original derivation of [2]. The resulting IV curve is shown in Fig. 6.8a and is in good agreement with [2].

The inclusion of a further decoherence mechanism through $\Delta$ is, in fact, not necessary in ECEID. By setting $\Delta = 0$, we see in Fig. 6.8b that all simulations converge well to the Landauer limit, even the case with the shortest leads $N_L = 40$ sites and $\Gamma = 0.75$ eV. $\Delta$ will not be included in any calculation in the rest of this work.

## 6.4  Joule Heating

We now apply the ECEID method to the conduction problem in a perfect nanowire, where phonons are inserted as Einstein oscillators in the central region, as shown in Fig. 6.9.

We simulate a perfect chain with leads made of 40 sites and a central region with 22 sites. All nearest-neighbour hoppings are 1 eV, onsite energies zero and $\Gamma = 0.75$ eV. A variable number of phonons is included in the central region, from none to 10, with a regular spacing. The electron-phonon coupling has the form

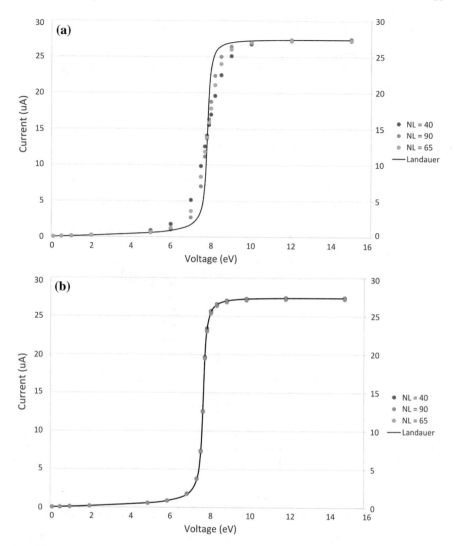

**Fig. 6.8** IV curves of the resonant dimer produced with ECEID. In **a** a one-to-one comparison with Fig. 6.7 (a finite $\Delta$ is included). In **b**, the same calculations as in (**a**) with $\Delta = 0$. The two-terminal Landauer result is shown as a solid black line

$$\hat{F}_v = F_v\Big(|n_v + 1\rangle\langle n_v| + |n_v\rangle\langle n_v + 1| - |n_v - 1\rangle\langle n_v| - |n_v\rangle\langle n_v - 1|\Big). \quad (6.6)$$

where $n_v$ is the site where phonon $v$ is located.[4] Every oscillator has $M = 0.5$ amu, $\hbar\omega = 0.2$ eV, $F = 1$ eV/Å and $N(t) = 0$. The oscillators are kept frozen at their

---

[4] To understand this form of $\hat{F}_v$, let us consider a perfect chain made of 3 electronic sites, described by Hamiltonian

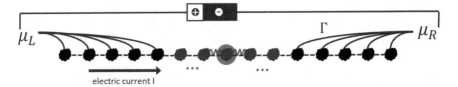

**Fig. 6.9** Schematic of a nanowire setup used to test Joule heating. The central region includes phonons treated as Einstein oscillators. Probes kept at chemical potentials $\mu_{L/R}$ are attached to the left/right leads to drive a current through the system

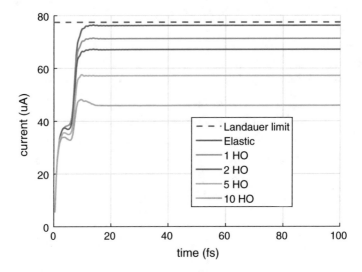

**Fig. 6.10** Dynamics of the current in ECEID simulations on a nanowire with an increasing number of harmonic oscillators, from none (elastic) up to 10. In these simulations, the phonons start at $N(0) = 0$ and are frozen with $\dot{N}(t) = 0$. The two-terminal Landauer limit is indicated with a dashed black line

initial condition (the ground state) by setting $\dot{N}(t) = 0$ for all times. As a result,

$$\hat{H} = \begin{bmatrix} 0 & t & 0 \\ t & 0 & t \\ 0 & t & 0 \end{bmatrix}.$$

The presence of an oscillator on site 2 would intuitively change the hoppings $t$ as a function of the oscillator displacement $\delta x$. A displacement towards one site corresponds to an equal and opposite in sign displacement towards the other site, so that the variation in the Hamiltonian caused by the oscillator can be written as

$$\delta \hat{H} = \begin{bmatrix} 0 & \frac{\delta t}{\delta x} & 0 \\ \frac{\delta t}{\delta x} & 0 & -\frac{\delta t}{\delta x} \\ 0 & -\frac{\delta t}{\delta x} & 0 \end{bmatrix} \delta x.$$

Defining $\hat{F} = -\frac{\delta \hat{H}}{\delta x}$, $F = \frac{\delta t}{\delta x}$ must be true, hence it is easy to recover the general case in Eq. (6.6).

they act as a perfectly efficient heat bath for the electrons. The electrons start at their ground state with a half-filled band. The current dynamics with the application of a bias $V = 1$ V is shown in Fig. 6.10 for a varying number of oscillators. It is remarkable to notice that the in this case the electrons are scattered by the zero point motion of the oscillators.

After a short transient of about 10 fs, the current in all ECEID simulations reaches a steady state. The case with no oscillators does not present any inelastic mechanism and reaches a steady state current value very close to the upper limit of the Landauer two-terminal case.[5] The introduction of the phonons introduces inelastic scattering for the electrons flowing in the nanowire, with the effect of reducing the steady state current. As intuitively expected, an increasing number of oscillators progressively decreases the final current value.

We now let $N(t)$ evolve in time, repeat the same set of simulations and show the current in Fig. 6.11a and the average phonon occupation $N_A(t) = \sum_v^{N_0} N_v(t)/N_0$ in (b). The time required for the current to reach a steady state increases because of the presence of the phonons that take a finite time to equilibrate during the process. When a current flows in the nanowire, the initially cold phonons heat up and, in turn, increase the scattering rate that the electrons perceive. This lowers the current that tends to reach a lower steady state value than the case with frozen phonons. This is an example of Joule heating and ECEID captures it.

To investigate the reduction in current in a more quantitative way, a slightly different setup needs to be devised.

### 6.4.1  A Microscopic Ohm's Law

We can picture an electron moving in a conductor as a particle that moves semi-classically between scattering events. In a perfect nanowire the electron does not encounter any scatterer and its motion is ballistic. When the density of scatterers is low, the system is in the diffusive regime and the resistance of a nanowire is given by Ohm's law [3]

$$R = r_0 \left( 1 + \frac{L}{l_0} \right), \tag{6.7}$$

where $r_0 = \hbar\pi/e^2 = 0.0129 V\mu$ A is the resistance quantum,[6] $l_0$ the electron mean free path (EMFP) and $L$ the length of the wire, in units of electronic sites.

The EMFP caused by a scattering potential $\hat{V}$ can be estimated perturbatively by Fermi's golden rule (FGR)

$$\frac{1}{l_0} = \frac{2\pi}{\hbar} |\langle i|\hat{V}|f\rangle|^2 \frac{d_f}{v} \tag{6.8}$$

---

[5]The values do not coincide because of finite size effects. In the simulations, the leads are finite and the two-terminal limit is approached by infinitely long leads.

[6]All the resistance results here will be given in terms of $r_0$.

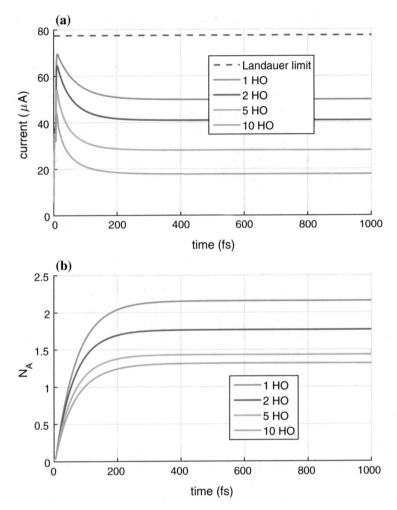

**Fig. 6.11** The same simulations from Fig. 6.10, with $N(t)$ allowed to change. In **a**, we show the current dynamics and in **b** the time evolution of $N_A(t)$

where $|\text{i}\rangle$ and $|\text{f}\rangle$ are the initial and final electronic state,[7] $d_f$ is the final density of states and $v$ is the electron's velocity. In the case of inelastic scattering due to a phonon $\nu$ from ECEID's model Hamiltonian (5.1), the scattering potential is $\hat{F}_\nu \hat{X}_\nu$.

We consider a perfect nanowire whose central region features a variable length with a constant oscillator density $n$. The phonons are identical Einstein oscillators and are localized periodically in the chain on sites $n_\nu$, with a $\hat{F}_\nu$ analogous to the one in Eq. (6.6). They all have mass $M$, characteristic frequency $\omega$ and are kept frozen by setting $\dot{N}_\nu(t) = 0$. Their occupation $N_\nu(0) = N(0)$ determines a collective temperature of

---

[7]In a wire with a left to right bias, as in our case, they are left and right propagating plane waves.

$T = \left(\hbar\omega/k_B\right)\left(N(0) + 1/2\right)$. The inelastic EMFP in such a system is

$$\frac{1}{l_{inel}} = \frac{4F^2}{a^2} n \frac{k_B T}{M\omega^2} \tag{6.9}$$

where $a$ is the hopping in the chain.[8] A variable $N(t)$ would change the scattering rate in time and the inelastic EMFP. Higher occupations $N(t)$ correspond to higher temperatures and shorter EMFP $l_{inel}$. This supports the observation from last section where the presence of hot phonons increased the scattering in the nanowire.

We simulate a perfect chain with $N_L = N_R = 50$ sites, onsite energies zero, hoppings 1 eV, $\Gamma = 0.5$ eV and a fixed bias $V = 1$ V. The initial electronic density matrix corresponds to a half filled band at zero electronic temperature. The central region has a variable length $N_C$ with a constant density of oscillators of 1 oscillator every 3 electronic sites. The simulations include up to 60 oscillators. The oscillators have $F = 0.5$ eV/Å, $M = 0.5$ amu, $\hbar\omega = 0.2$ eV and a constant $T = 5159$ K (which corresponds to $N(0) = 1.723$). The inelastic EMFP from Eq. (6.9) is 32.3 sites. We run simulations with different central region lengths, obtain a steady state current and plot the resistance curve in Fig. 6.12.

The resistance as a function of central region length is remarkably linear[9] and can be fitted with Ohm's law (6.7) to derive the system's resistivity. By using the perturbative EMFP (6.9), it is possible to estimate the resistivity as $r_0/l_{inel} = 0.0310 r_0$. This perturbative value is about 2% larger than the result from ECEID. Therefore, in these simulations, ECEID not only verifies Ohm's law at a microscopic level, but also agrees closely with the expected perturbative resistivity. This non-trivial agreement is a compelling validation of the physical accuracy of the ECEID method and the OB implementation.

---

[8]To prove this, we start from the energy band in a nearest-neighbour perfect chain $E = -2t \cos\phi$, where $\phi$ is a dimensionless momentum. The group velocity is

$$v = \frac{1}{\hbar}\frac{dE}{d\phi} = \frac{2a}{\hbar}\sin\phi \ . \tag{6.10}$$

and the density of states is

$$d_f = \frac{N_e}{2\pi}\frac{d\phi}{dE} = \frac{N_e}{2\pi}\frac{1}{2a\sin\phi} \tag{6.11}$$

where $N_e$ is the number of electronic sites. By inserting these quantities in Eq. (6.8), we have

$$\frac{1}{l_0} = \frac{N_e}{4a^2 \sin^2\phi}|\langle i|\hat{V}|f\rangle|^2. \tag{6.12}$$

To evaluate the backscattering of the potential in the main text, we use initial state $|\phi\rangle = 1/\sqrt{N_e}\sum_s^{N_e} e^{i\phi s}|s\rangle$ and final state $|-\phi\rangle$, where $s$ spans the atomic basis. Hence, $|\langle i|\hat{V}|f\rangle|^2$ becomes

$$|\langle-\phi|\hat{F}|\phi\rangle|^2 N_o\langle X^2\rangle = \frac{16F^2}{N_e^2} N_o \frac{k_B T}{M\omega^2} \tag{6.13}$$

where $N_o$ is the number of oscillators. Using these quantities on Eq. (6.12), we obtain Eq. (6.9).

[9]The $R^2$ that measures the goodness of fit is indeed very close to 1, as indicated in the figure.

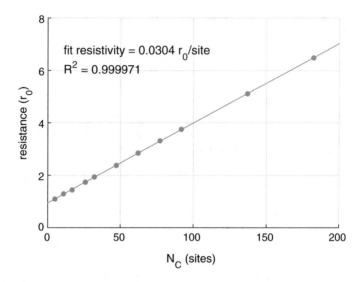

**Fig. 6.12** Wire resistance as a function of the length of the wire central region $N_C$. The density of oscillators $n$ is 1 oscillator every 3 electronic sites. The linear fit of the curve has a slope that corresponds to the nanowire resistivity. In the limit of a null central region, the ballistic Landauer result is recovered

### 6.4.2   Onsite Disorder

We introduce disorder in the central region of the chain to perform a test that includes both elastic and inelastic scattering. A disorder due to random onsite energies in the central region of a wire with an half-filled energy band produces the elastic EMFP[10]

$$\frac{1}{l_{\text{el}}} = \frac{A^2}{12a^2} \tag{6.15}$$

where $A$ is the amplitude of the disorder, with the onsite energies being sampled uniformly in the interval $[-A, A]$. When phonons and onsite disorder are present at the same time, the total EMFP is determined by both the elastic and inelastic scattering contributions:

$$\frac{1}{l_{\text{tot}}} = \frac{1}{l_{\text{inel}}} + \frac{1}{l_{\text{el}}}. \tag{6.16}$$

---

[10]To derive Eq. (6.15), one has to go back to Eq. (6.12) and use

$$|\langle -\phi|\hat{V}|\phi\rangle|^2 = |\sum_m^{N_e} E_m \langle -\phi|m\rangle\langle m|\phi\rangle|^2 = \frac{1}{N_e^2}\sum_m^{N_e} E_m^2 = \frac{1}{N_e}\langle \Delta E^2\rangle \tag{6.14}$$

where $\langle \Delta E^2\rangle$ is the variance of the onsite energies, $|m\rangle$ is the eigenstate basis of the chain and $|\phi\rangle = 1/\sqrt{N_e}\sum_s^{N_e} e^{i\phi s}|s\rangle$ is a plane wave on the atomic basis $|s\rangle$). By using $\langle \Delta E^2\rangle = A^2/3$ and imposing an half-filled band $\phi = \pi/2$, we obtain Eq. (6.15).

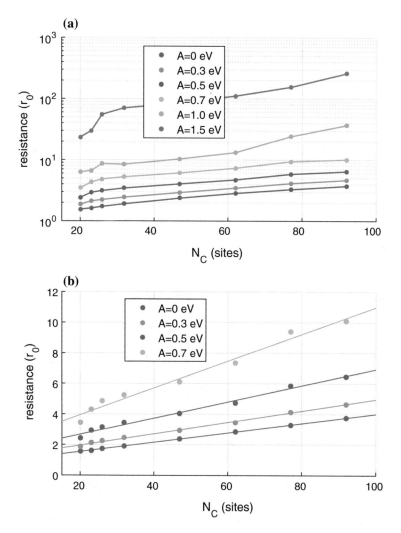

**Fig. 6.13** Panel (**a**): resistance (in logarithmic scale) as a function of $N_C$ in a nanowire such as the one from Sect. 6.4.1, where an onsite energy disorder is included. Panel (**b**): linear regression of the resistance for simulations with a disorder $A \leq 0.7$ eV. The fit slopes (resistivity) are shown in Table 6.1

We expect that, when $l_{inel} < l_{el}$, the dominant scattering mechanism is the inelastic one. In that case, the electrons are predominantly in the diffusive regime where the resistance is linear with the length of the central region. On the other hand, when $l_{el} < l_{inel}$, we expect the electrons dynamics to be mainly affected by the disorder. The electrons become localized and the resistance in the system grows exponentially with the length of the disordered region $N_C$.

**Table 6.1** Resistivity for disordered systems with an inelastic EMFP of $l_{inel} = 32.3$. The expected resistivity from the EMFP $l_{tot}$ is compared with the linear fit results from Fig. 6.13b

| A (eV) | $l_{el}$ (sites) | $l_{tot}$ (sites) | EMFP resistivity ($r_0$/site) | Fit resistivity ($r_0$/site) | Relative difference (%) |
|--------|------|------|------|------|------|
| 0.0 | ∞ | 32.3 | 0.0310 | 0.0304 | 1.9 |
| 0.3 | 120.2 | 25.4 | 0.0393 | 0.0371 | 5.7 |
| 0.5 | 48 | 19.3 | 0.0519 | 0.0512 | 1.2 |
| 0.7 | 24.5 | 13.7 | 0.0719 | 0.0835 | 13.9 |
| 1.0 | 12 | 8.7 | / | / | / |
| 1.5 | 5.3 | 4.6 | / | / | / |

We simulate the same configuration as the one in Sect. 6.4.1 and add a central region disorder up to $A = 1.5$ eV. The constant density of oscillators at a fixed temperature in the central region sets $l_{inel} = 32.3$. The interplay of the inelastic EMFP with the $l_{el}$ introduced by the onsite disorder determines the conduction regime of the system.

Figure 6.13a shows, as expected, a progressively increasing resistance with an growing disorder. The simulations with a low disorder $A = 0.3$ eV have a large $l_{el} = 120.2$ which makes $l_{tot} = 25$ quite close to the inelastic EMFP. The linear fit of the resistance curve in Fig. 6.13b gives a resistivity value that is in line with the one expected from the total EMFP. It is just 6% smaller than the EMFP resistivity, as we can see in Table 6.1. Simulations with a medium level of disorder $A = 0.5$ eV and $A = 0.7$ eV present an elastic EMFP that is comparable with the inelastic one and their total EMFP is, respectively, $l_{tot} = 19$ and $l_{tot} = 14$. Here the resistance regime is still linear and the resistivity from the fits is again in a good agreement with the expected one.

The cases with a higher disorder $A = 1.0$ eV and especially $A = 1.5$ eV start to deviate significantly from the linear regime, as is visible in Fig. 6.13a. There, the elastic scattering is the dominant mechanism and the resistance tends to grow exponentially with $N_C$. Transmission through highly disordered systems strongly depends on the configuration, therefore quantitative investigations of this regime require averaging over different configurations. Such a study could display the insurgency of Anderson localization [4].

## 6.5   Code Performance

We present performance tests to verify the efficiency and the scaling of the code with a varying number of electronic sites and oscillators.

To achieve a good scaling for large systems is a very difficult endeavour. Small details are enough to break parallelism in unexpected and difficult ways to fix. For example, during the code development, there was a puzzling situation where the code parallelism was broken. A pronounced slowdown would appear when the number of oscillators was close or equal to the number of cores of a machine's processor. The code ran on a server with 2 physical processors and 8 cores each. When executing the code with a few oscillators (up to about 6), the computational time was roughly independent of the oscillators number, as expected. When using 8 oscillators, the code would unexpectedly become several times slower. For 9–14 oscillators, the code would scale quite well, while for 16 oscillators it would get significantly slower again.

The problem lay in the update process between timesteps. The variable update involved fast data rewriting that mainly took place in the processor's cache. When the number of oscillators matched the number of cores in a physical processor, the size of the cache would become a bottleneck, as it was not enough to accommodate and process all the data at the same time. This turned the program into serial during the variable update stage, making the total computational time considerably longer. This issue was solved by redesigning the variable update process with pointers, a feature of modern Fortran. In *ElPh*'s current version the variable update does not involve overwriting data anymore, it involves switching labels of variables at different timesteps.

The code was executed on a computer with 20 cores, 10 each from 2 Intel Xeon E5-2680v2 CPUs, and 64 GB RAM. A number of simulations was performed on a nanowire with a geometry similar to the one used in the Joule heating tests in Sect. 6.4, with a number of electronic sites ranging from 50 to 500 and a number of oscillators from 1 to 40. The computational time of simulations with $10^6$ timesteps is shown in Table 6.2. The table also shows the scaling when compared to single oscillator simulations (oscillator scaling) and 50 electronic sites simulations (electronic scaling).

In the case of 50 electronic sites, the computational time is almost constant when going from 1 to 2 and 5 oscillators, a sign of a linear computational cost with the number of oscillators and of efficient parallelism. For 10–20 oscillators, the computational time tends to increase up to about twice the single oscillator performance. This can be explained by the overhead introduced by the large number of oscillators. The scaling in those cases is acceptable while, in the case of 40 oscillators, it is quite poor.

The parallelism is most efficient when the number of oscillators is less than the number of cores, nevertheless the code can still work with more oscillators than cores. In the case of the number of oscillators being twice the number cores, the scheduler of the operating system assigns on average the evolution of 2 oscillators to every core, making the code in principle twice as slow when compared to a code with half the number of oscillators. This procedure can be quite inefficient, especially for small systems such as the one with 50 electronic sites. In small systems, the load for every timestep is small and any overhead can have a large negative impact on the final computational time, hampering the code's efficiency and parallelism.

When increasing the number of electronic sites up to 500, the oscillator scaling from 1 to 5 oscillators remains remarkably quite linear. The overhead in the case for

**Table 6.2** Efficiency test of the code *ElPh* with dynamical simulations of systems with a variable number of electronic sites and oscillators including $10^6$ timesteps (which correspond to 1 ps when the timestep is 1 as, for example). The total computational times are shown, together with the oscillator scaling when compared to the single oscillator case and the electronic scaling when compared to the case with 50 electronic sites. Colours give a visual indication of the quality of the scaling: green stands for ideal (or close to), yellow for acceptable and red for poor scaling. In the oscillator scaling, performances less than two times the single core ($2\times$) are considered good, less than $4\times$ acceptable and beyond that poor. Because of the sparse matrix routines, the ideal electronic scaling would be quadratic with the number of electronic sites, i.e. $4\times$ for 100 sites, $9\times$ for 150 sites, $16\times$ for 200 sites and $100\times$ for 500 sites, when compared to the 50 sites case. There is a good scaling when the computational time is less than 1.5 times the ideal quadratic scaling, an acceptable scaling for less than 2.5 times and a poor scaling beyond that

| Number of electronic sites | Number of oscillators | Computational time | Oscillator scaling | Electronic scaling |
|---|---|---|---|---|
| 50 | 1 | 3 m 11 s | | |
| | 2 | 3 m 26 s | 1.1× | |
| | 5 | 3 m 58 s | 1.2× | |
| | 10 | 5 m 2 s | 1.6× | |
| | 15 | 5 m 46 s | 1.8× | |
| | 20 | 7 m 12 s | 2.3× | |
| | 40 | 33 m 49 s | 10.6× | |
| 100 | 1 | 13 m 8 s | | 4.1× |
| | 2 | 14 m 0 s | 1.1× | 4.1× |
| | 5 | 16 m 10 s | 1.2× | 4.1× |
| | 10 | 20 m 40 s | 1.6× | 4.1× |
| | 15 | 28 m 30 s | 2.2× | 4.9× |
| | 20 | 45 m 40 s | 3.5× | 6.3× |
| | 40 | 2 h 4 m 15 s | 9.5× | 3.7× |
| 150 | 1 | 31 m 9 s | | 9.8× |
| | 2 | 32 m 33 s | 1.0× | 9.5× |
| | 5 | 40 m 40 s | 1.3× | 10.3× |
| | 10 | 1 h 13 m 10 s | 2.3× | 14.5× |
| | 15 | 1 h 56 m 41 s | 3.7× | 20.2× |
| | 20 | 2 h 25 m 20 s | 4.7× | 20.2× |
| | 40 | 4 h 44 m 2 s | 9.1× | 8.4× |
| 200 | 1 | 1 h 0 m 20 s | | 19.0× |
| | 2 | 1 h 5 m 28 s | 1.1× | 19.1× |
| | 5 | 1 h 37 m 39 s | 1.6× | 24.6× |
| | 10 | 2 h 38 m 20 s | 2.6× | 31.5× |
| | 15 | 3 h 40 m 3 s | 3.6× | 38.2× |
| | 20 | 5 h 11 m 30 s | 5.2× | 43.3× |
| | 40 | 9 h 37 m 9 s | 9.6× | 17.1× |

(continued)

**Table 6.1** (continued)

| Number of electronic sites | Number of oscillators | Computational time | Oscillator scaling | Electronic scaling |
|---|---|---|---|---|
| 500 | 1 | 9h 55m 59s | | 187.2 |
| | 2 | 10h 37m 40s | 1.1 | 185.7 |
| | 5 | 15h 24m 30s | 1.6 | 233.1 |
| | 10 | 20h 16m 50s | 2.0× | 241.8× |
| | 15 | 1d 12h 35m 0s | 3.7× | 380.7× |
| | 20 | 2d 0h 37m 0s | 4.9× | 405.1× |
| | 40 | 3d 16h 46m 0s | 8.9× | 157.5× |

20 oscillators becomes more and more time consuming, making the code about 4 times slower than the single core performance, whereas the 40 oscillators case tends to take twice the time of the 20 oscillator case.

Focusing now on the electronic scaling, we effectively test the optimization in the code. A code without optimizations would scale about cubically with the number of electronic sites, while ideally it should scale quadratically. We cannot expect such a quadratic scaling though, especially for large systems, because there is at least one matrix operation that cannot be optimized to quadratic scaling: the term[11] $\hat{\rho}_e(t)\hat{F}_v\hat{\rho}_e(t)$ introduced in Sect. 5.5.

Going from 50 to 100 electronic sites, the code scales impeccably: in almost all the oscillator simulations the computational time scales quadratically and is 4 times higher. Considering the 150 electronic sites case, the electronic scaling is quadratic for up to 5 oscillators and it slightly increases up to twice that scaling for 20 oscillators. The most computationally demanding case is the one with 500 electronic sites. There, the electronic scaling is not quadratic any more, but it is not as poor as cubic either. In that case, the large size of the matrices makes the few operations that don't scale quadratically considerably more expensive and dominant on the final computational time.

The code can be furthermore optimized to improve performance and reduce overhead. Anyhow, the central aim of achieving a good oscillator scaling is fulfilled by the current version of the code. In Chap. 9 a more general version of the method is proposed, ECEID xp. In a well defined particular case, that method is equivalent to ECEID. Its implementation in the code is about 40% more efficient to simulate, as shown in Table 9.1, where the same tests as the ones presented here are performed.

---

[11] The whole operation $\hat{\rho}_e(t)\hat{F}_v\hat{\rho}_e(t)$ can be split in a sparse matrix multiplication first $\hat{F}_v\hat{\rho}_e(t) = \hat{G}_v(t)$ followed by a non-sparse multiplication $\hat{\rho}_e(t)\hat{G}_v(t)$. The last operation can still be optimized, as, if $\hat{F}_v$ is ultra sparse (e.g. Eq. (6.6)), $\hat{G}_v(t)$ is still quite sparse. For a more general $\hat{F}_v$ (e.g. Sect. (8.2.1)), the overall cost of the operation rises considerably. Anyhow, this multiplication is the slowest single operation in the code.

# References

1. Zrenner, A., E. Beham, S. Stufler, F. Findeis, M. Bichler, and G. Abstreiter. 2002. Coherent properties of a two-level system based on a quantum-dot photodiode. *Nature* 418 (6898): 612–614. https://doi.org/10.1038/nature00912.
2. McEniry, E.J., R. Bowler, D. Dundas, A.P. Horsfield, C.G. Sánchez, and T.N. Todorov. 2007. Dynamical simulation of inelastic quantum transport. *Journal of Physics: Condensed Matter*, 19 (19): 196201. https://doi.org/10.1088/0953-8984/19/19/196201
3. Todorov, T. 1996. Calculation of the residual resistivity of three-dimensional quantum wires. *Physical Review B–Condensed Matter and Materials Physics* 54 (8): 5801–5813. https://doi.org/10.1103/PhysRevB.54.5801
4. Anderson, P.W. 1958. Absence of diffusion in certain random lattices. *Physical Review* 109 (5): 1492–1505. https://doi.org/10.1103/PhysRev.109.1492.

# Chapter 7
# Thermalization with ECEID

Thermalization between electronic and vibrational degrees of freedom arises in a range of physical situations spanning widely different time and length scales. Examples were given in Chap. 2: they include Joule heating and dissipation in solid state and molecular physics [1, 2], equilibration of warm dense matter generated by laser pulses [3–5] and radiation cascades [6]. The interest in coupled dynamics of out-of-equilibrium electrons with vibrations occurs in several fields, including transport in molecular junctions [7, 8] and photoelectron spectroscopy [9], and has triggered the development of new experimental techniques [10].

Meanwhile, real-time atomistic simulations venture more and more often into non-equilibrium problems where accounting for electron-phonon thermalization is crucial [11]. A choice of methods can capture the interaction between electrons and vibrations, from the phenomenological Boltzmann equation in extended systems [12] to its counter-part at the nanoscale, non-equilibrium Green's functions (NEGF) [13].

Nevertheless, the problem of thermal equilibration between interacting degrees of freedom (DOF) is particularly difficult to tackle from the simulation point of view. For purely classical systems simulated via Molecular Dynamics, anharmonicities in the potential can lead to thermalization and energy equipartition [14]. In harmonic or weakly anharmonic systems, equilibration does not happen spontaneously: it requires the introduction of external thermostats.

The situation is even more complicated for quantum interacting systems and it becomes especially critical in mixed quantum-classical approaches. A widely used approach is the macroscopic two-temperature model that we described in Chap. 3. In that model, the nuclear and the electronic motion are represented in terms of temperature fields coupled via appropriate diffusion equations [15, 16]. This together with the introduction of Langevin thermostats [17] has proved successful in interpreting measured quantities [18, 19]. This approach remains of active interest and, in recent years, it has evolved into more elaborate methodologies where the nuclear motion is taken into account via classical molecular dynamics simulations while electrons are treated at increasing levels of sophistication [20–25].

© Springer International Publishing AG, part of Springer Nature 2018
V. Rizzi, *Real-Time Quantum Dynamics of Electron-Phonon Systems*,
Springer Theses, https://doi.org/10.1007/978-3-319-96280-1_7

We described non-adiabatic electron-nuclear atomistic simulations in Chap. 4. The simplest non-adiabatic approach is Ehrenfest dynamics (ED) [1] in which classical nuclei interact with the mean electron density. ED is tractable and simple but it fails to describe the spontaneous decay of electronic excitations into phonons because of the lack of microscopic detail in the electronic density and resultant loss of electron-nuclear correlation [26]. Vibrational DOF spontaneously cool down at the expense of increasing the electronic energy, violating the second law of thermodynamics. What is missing in ED are the collisions that drive the probability distribution function towards equilibrium. The approach to equilibrium can be reinstated via Boltzmann's kinetic theory, i.e. through phenomenological relaxation dynamics. However to recover this in microscopic dynamics for a closed system requires thermostating techniques; for quantum DOF this introduces an additional layer of complexity.

Correlated electron-ion dynamics (CEID) [26, 27] is a method that was developed to go beyond ED. It starts from the bare electron-nuclear Hamiltonian and solves it approximately by a perturbative expansion in powers of nuclear fluctuations about the mean trajectory. However it scales between quadratically and cubically with the number of nuclear DOF, becoming prohibitive beyond a few DOF, along with difficulties in the choice of closure strategy for the hierarchy of perturbative equations of motion. The computational bottleneck persists in alternative expansion strategies for the electron-nuclear problem [28].

Today there is a new impetus in the study of mesoscale systems, as their technological applications and simulation capability meet [29]. These systems mark a difficult middle ground between bulk and the atomic scale. There is a serious need for a methodology that includes the mechanisms of thermal equilibration between electron and phonon DOF, and at the same time is amenable to computer simulation with present day resources [30]. This need for an efficient approach to the dynamics of thermalization at the mesoscale has motivated the development of a microscopic method for coupled real-time quantum electron-phonon dynamics, ECEID, that is described in Chap. 5.

ECEID advances beyond CEID in terms of conceptual and computational tractability by exploiting a different starting point: a system of electrons and harmonic vibrations, coupled by an interaction linear in the generalized displacements. This more specialized scenario maintains applicability to the large family of problems involving harmonic nuclear motion, while offering important advantages. This Hamiltonian starts from the Born-Oppenheimer level of description, with the role of the coupling being to generate the non-adiabatic corrections. By contrast, the old CEID method above had the dual challenge of first generating the Born-Oppenheimer behaviour (starting from the bare full Hamiltonian), and then also going beyond. Furthermore ECEID employs a non-perturbative closure strategy, which enables the coupled electron-phonon dynamics to be formulated in terms of a set of variables and equations of motion that scale linearly with the number of vibrational DOF. This opens the possibility of tackling problems previously out of reach: as we showed in Sect. 6.5, in test runs we have been able to simulate up to 500 electrons interacting with 40 vibrational DOF on the picosecond time-scale, on an ordinary workstation.

The contents in this chapter have been presented in [31].

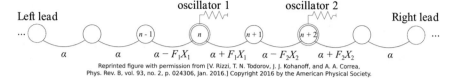

**Fig. 7.1** Schematic of our model system: a nearest-neighbour one-dimensional lattice model of
an atomic wire divided into a central region between two leads. This embeds the sample in an
environment. Each of the 3 regions has 32 sites, with 15 equispaced harmonic oscillators coupled
to the central region. Oscillator $\nu$ couples to site $n_\nu$ through $\hat{F}_\nu = F_\nu\left(\hat{c}^\dagger_{n_\nu+1}\hat{c}_{n_\nu} + \hat{c}^\dagger_{n_\nu}\hat{c}_{n_\nu+1} - \hat{c}^\dagger_{n_\nu}\hat{c}_{n_\nu-1} - \hat{c}^\dagger_{n_\nu-1}\hat{c}_{n_\nu}\right)$ which corresponds to independent atomic motion in a lattice description.
The extension from Einstein oscillators to normal modes is straightforward. The onsite energies
are uniform, the hoppings $\alpha = -1$ eV and $\Gamma = 0.08$ eV. For all the oscillators $M = 0.5$ a.m.u.,
$\hbar\omega = 0.2$ eV and $F = 0.3$ eV/Å [31]

## 7.1 The System

Here we have implemented the ECEID method for the discretized electron-phonon
Hamiltonian (7.1)

$$\hat{H}_{e-ph} = \overbrace{\sum_{ij}\alpha_{ij}\hat{c}^\dagger_i\hat{c}_j}^{\hat{H}_e} - \sum_{\nu ij}F_{\nu ij}\hat{c}^\dagger_i\hat{c}_j\overbrace{\frac{\hat{a}^\dagger_\nu + \hat{a}_\nu}{\sqrt{2M_\nu\omega_\nu/\hbar}}}^{\hat{F}_\nu\hat{X}_\nu} + \sum_\nu\hbar\omega_\nu\left(\hat{a}^\dagger_\nu\hat{a}_\nu + \frac{1}{2}\right) \quad (7.1)$$

where $\hat{c}^\dagger(\hat{c})$ are the fermion creation (annihilation) operators. $\alpha_{ij}$ are onsite energies
and hoppings with $\{i, j\}$ running over the atomic sites. The electronic DM evolves
according to Eq. (5.5). $\hat{\mu}_\nu(t)$ is calculated using Eq. (5.34), which is obtained from
the time evolution of the auxiliary operators (5.36–5.39). The Hamiltonian driving
those auxiliary operators includes of a small $\Gamma$, following Sect. (6.2), to mimic an
extended system without the extra cost. These quantities enter also in the EOM for
the mean oscillator occupation (5.41). The number of EOM scales linearly with $N_o$
and so does the computational cost.

We use these equations to simulate non-equilibrium electron-phonon dynamics
in the model in Fig. 7.1: a wire with an electronic half-filled band with 96 spin-
degenerate non-interacting electrons coupled to 15 harmonic oscillators.

## 7.2 An Entropic Definition of Temperature

To track the evolution of the two subsystems, we use two temperature-like parame-
ters: $T_0^{quant}$ for the oscillators and $T_e$ for electrons. If $\overline{N}(t) = \sum_{\nu=1}^{N_o}N_\nu(t)/N_o$, then
the oscillator temperature is defined through $\overline{N}(t) = (e^{\hbar\omega/k_B T_0^{quant}(t)} - 1)^{-1}$. In the

Ehrenfest case, this definition breaks down when the energy of the classical oscillators goes down to zero and $\overline{N}(t) \to -1/2$. For that case, we employed an alternative semiclassical definition of oscillator temperature $k_B T_0^{\text{class}} = (\overline{N}(t) + \frac{1}{2})\hbar\omega$. The electronic temperature is taken from $T_e = \Delta E_e / \Delta S_e$ where $\Delta E_e$ is the variation over 5 timesteps in electronic energy and $\Delta S_e$ is the corresponding variation in electronic von Neumann entropy $S_e = -k_B \sum_n (f_n \log f_n + (1 - f_n) \log(1 - f_n))$, where $f_n$ are the diagonal elements of $\hat{\rho}_e$ in the basis of $\hat{H}_e$ eigenstates, the occupations of the unperturbed electronic energy levels. $T_e$ is then inferred from a running average of its reciprocal. We note that these temperatures are only observables, not an input into the simulation.

As the system evolves, no macroscopic work is done, but energy (heat) is exchanged between the electronic and the oscillators subsystems. Having a microscopic definition of the entropy also allows us to give a time-local quantification of the rate of heat exchange $J_Q = \frac{dS_{\text{total}}}{dt}/(1/T_0 - 1/T_e)$, where $dS_{\text{total}} = dS_0 + dS_e$. In the weak-coupling limit, where the correlation energy $E_c$ is small, the heat current reduces to $J_Q = dE_0/dt$, and on average $dE_0/dt = -dE_e/dt$.

## 7.3  Comparison with Ehrenfest Dynamics

The terms involving $[\hat{F}_\nu, \hat{\rho}_e(t)]$ in Eq. (5.36) are related to the electronic friction (an effective dissipative force due to electron-hole excitations by the oscillator), while those with $\{\hat{F}_\nu, \hat{\rho}_e(t)\}$ in Eq. (5.38) describe electronic noise and spontaneous phonon emission [32, 33].

To see this, consider the above problem within Ehrenfest dynamics: electrons interacting with a classical oscillator, with phase $\phi$, slowly varying amplitude $A$, displacement $X(t) = A \sin(\omega t - \phi)$, and velocity $V(t) = \dot{X}(t)$. Next, average over $\phi$, to sample different initial conditions. The counterpart of the earlier approximations reads $\langle X(t)X(\tau)\hat{\rho}_e(\tau, \phi)\rangle_\phi \approx \langle X(t)X(\tau)\rangle_\phi \hat{\rho}_e(\tau)$, together with suppression of oscillator position-momentum correlations. This produces (5.34) *without* the second term, and with $N$ given by $(N + 1/2)\hbar\omega = M\omega^2 A^2/2$.

The phase-averaged power into the Ehrenfest oscillator, $\langle V(t)F(t)\rangle_\phi$ with $F(t) = \text{Tr}_e(\hat{F}\hat{\rho}_e(t, \phi))$, becomes (5.41)

$$\dot{N}(t) = \frac{1}{M\hbar\omega}\left(\text{i}\,\text{Tr}_e(\hat{F}\hat{C}^s(t)) + \text{Tr}_e(\hat{F}\hat{A}^c(t))\right), \qquad (7.2)$$

without the second term. Finally, the remaining first term in (7.2) is the same as the mean rate of work by the electronic friction force due to the symmetric part of the velocity-dependent force kernel in Eq. (16) in [33]. Thus the ECEID EOM with the anticommutator in (5.38) suppressed describe ED (with oscillator phase averaged out), physically dominated by electron-hole excitations and electronic friction.

The second term in (7.2) corresponds instead to the power delivered to the oscillators by the effective electronic-noise force described by line 1 of Eq. (56) in [33]: the

key correction beyond the mean-field ED. The competition between the two terms in
(7.2) enables thermodynamic electron-phonon equilibration [33], which is thus built
into the ECEID method.

## 7.4 Results

### 7.4.1 Thermalization

Our first example starts with $T_e = 10000$ K and $T_0^{class} = 1400$ K. This mimics a
common situation in laser or irradiation experiments in which electrons initially
absorb energy faster than ions [34]. In Fig. 7.2 we compare the time evolution of the
temperature for ED and ECEID. After a short transient which depends on the details
of the initial state, a long-lived steady state develops with a net energy flow from
one subsystem to the other. In ED, the absence of electronic noise (second term in
Eq. 7.2) results in a heat flow going in the wrong direction: from the cold oscillators
into the hot electrons, until the oscillators reach zero temperature. In ECEID, the
inclusion of the electronic noise makes the exchange of heat physical and the final
thermalization possible (Fig. 7.2a). The heat flow scales linearly with the temperature
difference (Fourier's law) (Fig. 7.2c). In the equilibrium state reached in ECEID, the
two final temperatures agree within 1%.

### 7.4.2 Population Inversion

Next, we test an extremely out-of-equilibrium phenomenon: a complete population
inversion. Initially, the electrons occupy the upper half of the energy states in the
wire, corresponding to an infinitesimal negative electronic temperature. The oscillators are held at $N = 0.5$, or $T_0^{quant} = 2112$ K throughout. This simulates coupling to
an infinitely efficient external thermostat, thus isolating just the electron dynamics.
Figure 7.3 shows snaphots of the electronic population dynamics and the temperature.
The electrons de-excite in both ECEID and ED. In ED this happens through negative
friction [35]. Comparing Fig. 7.3a and b at 0.5 ps, we see that the de-excitation is
faster in ECEID; this is because ECEID includes also the contribution from spontaneous phonon emission. But the crucial difference is the final state: ECEID correctly
takes the electrons all the way down to a Fermi-Dirac distribution corresponding to
the oscillator temperature; ED by contrast gets stuck at a distribution with roughly
uniform occupancies [36]. These two ED features have a common origin. If electronic occupancies $f(E)$ depend only on energy $E$, then a rearrangement of the
result for the electronic friction in [8] gives an integral containing $f'(E)$ as a factor in the integrand. Hence the opposite signs for the friction, at small negative and
small positive temperatures. Hence also the unphysical "equilibration" of the elec-

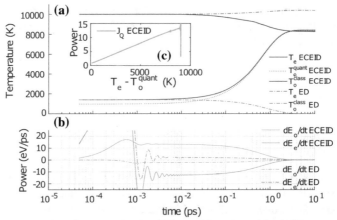

**Fig. 7.2** Coupled dynamics of a closed system of electrons and oscillators with the parameters given in the text. **a** Time evolution of the electronic and oscillators temperature for ECEID and the phase-averaged ED discussed above. **b** Rate of change of electronic and oscillators energies. After a transient of 10 fs, the systems evolve until eventually an equillibrium state (ECEID) or an unphysical state (ED) is reached. **c** For ECEID a clear linear scaling (Fourier law behavior) is observed for heat flow versus temperature difference (up to a time of 2.5 ps). The noise for high temperature differences is related to the initial transient [31]

trons at $f'(E) = 0$ in ED, when the friction vanishes and the main electron-phonon interaction mechanism present in ED goes to zero.

The role of $\Gamma$ in these simulations is crucial for thermalization because it provides a controlled way to embed a finite system, that would not equilibrate, into an extended one that does. In Fig. 7.4 we study the time evolution of a sample of electronic states in ECEID for $\Gamma = 0.08$ eV and $\Gamma = 0.8$ eV for the same initial population inversion as above. The results are almost superimposable: for $\Gamma$ larger than the average level spacing $\sim 0.04$ eV, ECEID is largely independent of $\Gamma$. We observed that the dynamics of any level $j$ is exactly symmetric with that of level $96 - j + 1$ for all times.

### 7.4.3  Kinetic Model

The rich pattern of population evolutions shown in Fig. 7.4 can be understood with a kinetic model of the transitions between electronic levels due to phonon absorption and emission. The rate equation for the population $f_j$ of level $j$ is

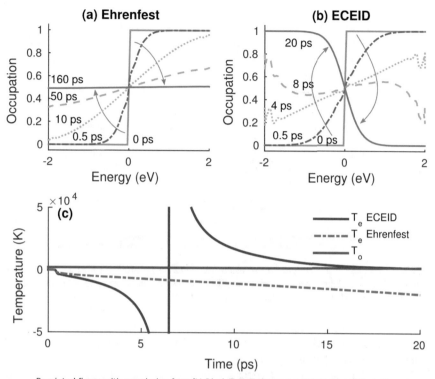

**Fig. 7.3** Population inversion simulation with the oscillators held at constant temperature. We show snapshots of the population of the electronic states in **a** ED at 0 ps, 0.5 ps, 10 ps, 50 ps, 160 ps and **b** ECEID at 0 ps, 0.5 ps, 4 ps, 8 ps, 20 ps. (The arrows highlight the overall initial-to-final transition in each case.) **c** Temperature evolution during the simulation for ED and ECEID compared with the fixed oscillator temperature [31]

$$\dot{f}_j(t) = \sum_k \frac{1}{\tau_{jk}} (-Nf_j(1 - f_k) + (N + 1)f_k(1 - f_j))$$

$$+ \frac{1}{\tau_{kj}} (Nf_k(1 - f_j) - (N + 1)f_j(1 - f_k)). \tag{7.3}$$

The scattering rates $1/\tau_{jk} = (\pi/M\omega)N_0|F_{jk}|^2 G_{jk}$ are given by the Fermi Golden Rule (FGR). $|F_{jk}|^2$ can be calculated analytically by using plane wave states with energies $E_j = 2\alpha \cos \phi_j$, (dimensionless) crystal momentum $\phi_j = j\pi/97$, $j = 1, \ldots, 96$ and by averaging over the two opposite signs of momentum for the final state. $G_{jk} = e^{-((E_k - E_j - \hbar\omega)/\Delta)^2}/(\sqrt{\pi}\Delta)$ is a Gaussian envelope with a width $\Delta$. It mimics the $\delta$-function that appears in the FGR electron-phonon transition rates. We plug the parameters of the population inversion simulation from Fig. 7.3 into

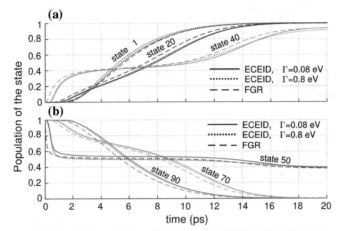

**Fig. 7.4** Comparison of the dynamics of electronic states for ECEID with $\Gamma = 0.08$ eV, ECEID with $\Gamma = 0.8$ eV and the kinetic model starting from an inverted population. In **a** we track state 1, state 20 and state 40; in **b** state 90, state 70 and state 50 [31]

the kinetic model with $\Delta = 0.08$ eV and in Fig. 7.4 we compare it with ECEID simulations, showing close agreement.

The comparison with the kinetic model illustrates that ECEID, owing to its scalability, can access time- and size-domains where macroscopic thermodynamic behaviour is beginning to emerge. In addition, the direct comparison between inherently different descriptions provides a bottom-up path to a validation, at the atomistic level, of kinetic models of electron-phonon dynamics, without having to resort to the relaxation-time approximation.

The response is fastest for the states in the middle of the band, i.e. the states close to the step in the initial population. The time that these states take to settle into a long-lived half-occupied steady state—about 0.5 ps—is comparable to the time needed for the initial temperature response—the small initial step-like feature in the blue results in Fig. 7.3c. (This transient response in the electron-phonon dynamical simulation is absent in FGR, because FGR by construction describes mean transition rates in the long-time limit.) The results of the kinetic model show little variation over the range $0.04 < \Delta < 0.15$ eV or for different shapes of $G_{jk}$. For this choice of parameters, the kinetic model captures the main physics of the problem. The combination of the kinetic model and ECEID provides a direct way to construct rate equations that allow thermodynamic electron-phonon equilibration on the basis of a real-time quantum mechanical simulation. The low computational cost of the kinetic model opens up the possibility to simulate large scale systems. Its application to irradiated metallic systems is currently under development.

# References

1. Horsfield, A.P., D.R. Bowler, H. Ness, C.G. Sánchez, T.N. Todorov, and A.J. Fisher. 2006. The transfer of energy between electrons and ions in solids. *Reports on Progress in Physics* 69 (4): 1195–1234. https://doi.org/10.1088/0034-4885/69/4/R05.
2. Galperin, M., M.A. Ratner, and A. Nitzan. 2007. Molecular transport junctions: Vibrational effects. *ournal of Physics: Condensed Matter* 19: 103201. https://doi.org/10.1088/0953-8984/19/10/103201.
3. Fann, W.S., R. Storz, H.W.K. Tom, and J. Bokor. 1992. Direct measurement of nonequilibrium electron-energy distributions in subpicosecond laser-heated gold films. *Physical Review Letters* 68 (18): 2834–2837. https://doi.org/10.1103/PhysRevLett.68.2834.
4. Fann, W.S., R. Storz, H.W.K. Tom, and J. Bokor. 1992. Electron thermalization in gold. *Physical Review B* 46 (20): 13 592–13 595. https://doi.org/10.1103/PhysRevB.46.13592.
5. Ogitsu, T., Y. Ping, A. Correa, B.I. Cho, P. Heimann, E. Schwegler, J. Cao, and G.W. Collins. 2012. Ballistic electron transport in non-equilibrium warm dense gold. *High Energy Density Physics* 8 (3): 303–306. https://doi.org/10.1016/j.hedp.2012.01.002.
6. Duffy, D., S. Khakshouri, and A. Rutherford. 2009. Electronic effects in radiation damage simulations. *Nuclear Instruments and Methods in Physics Research Section B: Beam Interactions with Materials and Atoms* 267 (18): 3050–3054. https://doi.org/10.1016/j.nimb.2009.06.047.
7. Härtle, R., and M. Thoss. 2011. Vibrational instabilities in resonant electron transport through single-molecule junctions. *Physical Review B* 83 (12): 125419. https://doi.org/10.1103/PhysRevB.83.125419.
8. Lü, J.-T.T., M. Brandbyge, P. Hedegård, T.N. Todorov, D. Dundas, P. Hedegard, T.N. Todorov, and D. Dundas. 2012. Current-induced atomic dynamics, instabilities, and Raman signals: Quasiclassical Langevin equation approach. *Physical Review B* 85 (24): 245444. https://doi.org/10.1103/PhysRevB.85.245444.
9. Avigo, I., R. Cortés, L. Rettig, S. Thirupathaiah, H.S. Jeevan, P. Gegenwart, T. Wolf, M. Ligges, M. Wolf, J. Fink, and U. Bovensiepen. 2013. Coherent excitations and electron phonon coupling in Ba/EuFe2As2 compounds investigated by femtosecond time- and angle-resolved photoemission spectroscopy. *Journal of Physics: Condensed Matter* 25 (9): 094003. https://doi.org/10.1088/0953-8984/25/9/094003.
10. Lewis, N.H.C., H. Dong, T.A.A. Oliver, and G.R. Fleming. 2015. Measuring correlated electronic and vibrational spectral dynamics using line shapes in two-dimensional electronic-vibrational spectroscopy. *The Journal of Chemical Physics* 142 (17): 174202. https://doi.org/10.1063/1.4919686.
11. Kogoj, J., L. Vidmar, M. Mierzejewski, S.A. Trugman, and J. Bonča. 2016. Thermalization after photoexcitation from the perspective of optical spectroscopy. *Physical Review B–Condensed Matter and Materials Physics* 94 (1): 014304. https://doi.org/10.1103/PhysRevB.94.014304.
12. Ashcroft, N.W., and D.N. Mermin. 1976. *Solid State Physics*. Philadelphia: Saunders College.
13. Frederiksen, T., M. Paulsson, M. Brandbyge, and A.-P. Jauho. 2007. Inelastic transport theory from first principles: Methodology and application to nanoscale devices. *Physical Review B* 75 (20): 205413. https://doi.org/10.1103/PhysRevB.75.205413.
14. Frenkel, D., and B. Smit. 2002. *Understanding molecular simulation*. Academic Press.
15. Anisimov, S.I., B.L. Kapeliovich, and T.L. Perel'man. 1975. Electron emission from metal surfaces exposed to ultrashort laser pulses. *Journal of Experimental and Theoretical Physics* 39: 375–377.
16. Flynn, C.P., and R.S. Averback. 1988. Electron-phonon interactions in energetic displacement cascades. *Physical Review B* 38 (10): 7118. https://doi.org/10.1103/PhysRevB.38.7118.
17. Caro, A., and M. Victoria. 1989. Ion-electron interaction in molecular-dynamics cascades. *Physical Review A* 40 (5): 2287–2291. https://doi.org/10.1103/PhysRevA.40.2287.
18. Finnis, M.W., P. Agnew, and A.J.E. Foreman. 1991. Thermal excitation of electrons in energetic displacement cascades. *Physical Review B* 44 (2): 567–574. https://doi.org/10.1103/PhysRevB.44.567.

19. Prönnecke, S., A. Caro, M. Victoria, T.D. de la Rubia, and M. Guinan. 1991. The effect of electronic energy loss on the dynamics of thermal spikes in Cu. *Journal of Materials Research* 6 (03): 483–491. https://doi.org/10.1557/JMR.1991.0483.
20. Duffy, D.M., and A.M. Rutherford. 2007. Including the effects of electronic stopping and electron-ion interactions in radiation damage simulations. *Journal of Physics: Condensed Matter* 19 (1): 016207. https://doi.org/10.1088/0953-8984/19/1/016207.
21. Race, C.P., D.R. Mason, M.W. Finnis, W.M.C. Foulkes, A.P. Horsfield, and A.P. Sutton. 2010. The treatment of electronic excitations in atomistic models of radiation damage in metals. *Reports on Progress in Physics* 73 (11): 116501. https://doi.org/10.1088/0034-4885/73/11/116501.
22. Mason, D. 2015. Incorporating non-adiabatic effects in embedded atom potentials for radiation damage cascade simulations. *Journal of Physics: Condensed Matter* 27 (14): 145401. https://doi.org/10.1088/0953-8984/27/14/145401.
23. Cho, B.I., K. Engelhorn, A.A. Correa, T. Ogitsu, C.P. Weber, H.J. Lee, J. Feng, P.A. Ni, Y. Ping, A.J. Nelson, D. Prendergast, R.W. Lee, R.W. Falcone, and P.A. Heimann. 2011. Electronic structure of warm dense copper studied by ultrafast X-ray absorption spectroscopy. *Physical Review Letters* 106 (16): 167601. https://doi.org/10.1103/PhysRevLett.106.167601.
24. Karim, E.T., M. Shugaev, C. Wu, Z. Lin, R.F. Hainsey, and L.V. Zhigilei. 2014. Atomistic simulation study of short pulse laser interactions with a metal target under conditions of spatial confinement by a transparent overlayer. *Journal of Applied Physics* 115 (18): 183501. https://doi.org/10.1063/1.4872245.
25. Zarkadoula, E., S.L. Daraszewicz, D.M. Duffy, M.A. Seaton, I.T. Todorov, K. Nordlund, M.T. Dove, and K. Trachenko. 2014. Electronic effects in high-energy radiation damage in iron. *Journal of Physics. Condensed Matter: An Institute of Physics Journal* 26 (8): 085401. https://doi.org/10.1088/0953-8984/26/8/085401.
26. Horsfield, A.P., D.R. Bowler, A.J. Fisher, T.N. Todorov, and C.G. Sánchez. 2004. Beyond ehrenfest: Correlated non-adiabatic molecular dynamics. *Journal of Physics: Condensed Matter* 16 (46): 8251–8266. https://doi.org/10.1088/0953-8984/16/46/012.
27. Horsfield, A.P., D.R. Bowler, A.J. Fisher, T.N. Todorov, and C.G. Sanchez. 2005. Correlated electron-ion dynamics: The excitation of atomic motion by energetic electrons. *Journal of Physics: Condensed Matter* 17 (30): 4793–4812. https://doi.org/10.1088/0953-8984/17/30/006.
28. Stella, L., M. Meister, A.J. Fisher, and A.P. Horsfield. 2007. Robust nonadiabatic molecular dynamics for metals and insulators. *The Journal of Chemical Physics* 127 (21): 214104. https://doi.org/10.1063/1.2801537.
29. Wang, L., R. Long, and O.V. Prezhdo. 2015. Time-domain Ab initio modeling of photoinduced dynamics at nanoscale interfaces. *Annual Review of Physical Chemistry* 66 (1): 549–579. https://doi.org/10.1146/annurev-physchem-040214-121359.
30. Cahill, D.G., P.V. Braun, G. Chen, D.R. Clarke, S. Fan, K.E. Goodson, P. Keblinski, W.P. King, G.D. Mahan, A. Majumdar, H.J. Maris, S.R. Phillpot, E. Pop, and L. Shi. 2014. Nanoscale thermal transport. II. 2003–2012. *Applied Physics Reviews* 1 (1): 011305.
31. Rizzi, V., T.N. Todorov, J.J. Kohanoff, and A.A. Correa. 2016. Electron-phonon thermalization in a scalable method for real-time quantum dynamics. *Physical Review B* 93 (2): 024306. https://doi.org/10.1103/PhysRevB.93.024306.
32. Mozyrsky, D., and I. Martin. 2002. Quantum-classical transition induced by electrical measurement. *Physical Review Letters* 89 (1): 018301. https://doi.org/10.1103/PhysRevLett.89.018301.
33. Todorov, T.N., D. Dundas, J.-T. Lü, M. Brandbyge, and P. Hedegard. 2014. Current-induced forces: a simple derivation. *European Journal of Physics* 35 (6): 065004. https://doi.org/10.1088/0143-0807/35/6/065004.
34. Lisowski, M., P. Loukakos, U. Bovensiepen, J. Stähler, C. Gahl, and M. Wolf. 2004. Ultrafast dynamics of electron thermalization, cooling and transport effects in Ru(001). *Applied Physics A: Materials Science & Processing* 78 (2): 165–176. https://doi.org/10.1007/s00339-003-2301-7.

35. Lü, J.-T., P. Hedegård, and M. Brandbyge. 2011. Laserlike vibrational instability in rectifying molecular conductors. *Physical Review Letters* 107 (4): 046801. https://doi.org/10.1103/PhysRevLett.107.046801.

36. Theilhaber, J. 1992. Ab initio simulations of sodium using time-dependent density-functional theory. *Physical Review B* 46 (20): 12 990–13 003. https://doi.org/10.1103/PhysRevB.46.12990.

# Chapter 8
# Inelastic Electron Injection in Water

When high-energy radiation penetrates living cells, it ionizes molecules along its path and can cause cell death by damaging DNA. Radiation events involve a sequence of processes that ultimately require a clear microscopic understanding [1]. Only about one third of the cellular damage is produced by direct interaction of the ionizing radiation with DNA, while the rest is due to secondary species, produced in the first hundreds to thousands of femtoseconds following the primary irradiation of the system [2].

Secondary electrons are a key irradiation by-product, as about $\simeq 50 \cdot 10^3$ electrons are emitted for every MeV of incoming energy [3]. The majority of secondary electrons are low energy electrons (LEEs), with an energy distribution peaking below 10 eV [4]. It may seem intuitive that the higher the electronic energy, the more significant the damage, and that electrons with energies below the DNA ionization threshold $\approx 15$ eV cannot cause strand breaks and destroy DNA, but this perception was challenged in 2000 [2, 5]. It was discovered that LEEs with energies between 3 and 20 eV can damage DNA considerably and their damaging power does not constantly increase with their energy. These studies triggered an intense effort into understanding the interaction mechanisms of LEEs with DNA [1, 3, 6–8].

The study of LEEs dynamics and their interaction with the cellular environment hinges on the non-adiabatic evolution of molecular systems over picosecond timescales. The mesoscale nature of this problem makes it especially challenging for non-adiabatic quantum electron-nuclear simulations. The ECEID method developed in Chap. 5 is designed for such simulations. Here we combine it with electronic Open Boundaries (OB), as in Sect. 5.7, to simulate in real time the injection of LEEs and their dynamical interaction with phonons. Part of the content in this chapter was published in Scientific Reports [9].

Water is the main component of cells: most LEEs are generated from it and, in turn, interact with it, while its presence plays an enhancing role in DNA radiation

© Springer International Publishing AG, part of Springer Nature 2018
V. Rizzi, *Real-Time Quantum Dynamics of Electron-Phonon Systems*,
Springer Theses, https://doi.org/10.1007/978-3-319-96280-1_8

damage [10]. Electron tunneling through static water configurations has been studied [11–13], focusing on resonance lifetimes and the relevance of inelastic effects at a perturbative level. We adopt a simple model of water and, at first, we test a water molecule with its two stretching modes. We compare ECEID with elastic results, investigate the high mass limit and apply a bias to the molecule letting a current flow through it.

Then we implement a water chain with one phonon mode to mimic a minimal biological environment and inject LEEs into it at different energies. Phonon absorption and emission play an essential role for enabling electrons to enter the water chain. Phonon-assisted injection shows a great sensitivity to vibrational temperature, with the possibility of dramatically reducing or enhancing the electron flow. Phonons are therefore a crucial control factor for the injection of LEEs into water. The excited states display an energy dependent lifetime that we compare with self-energy results and offer a possible electron trapping mechanism. In Appendix C, we include results about a simplified water chain system and the formation of a peculiar state with a much longer lifetime than the other states.

It is hoped that these results will provide a framework for further dynamical simulations of radiation damage in more complex biological systems [14, 15].

## 8.1  Water Molecule

We begin by applying ECEID to the elastic and inelastic transmission properties of a single water molecule.

### 8.1.1  A Simple Water Model

We require a simple model that is easy to implement and makes the inclusion of phonons intuitive. We choose a planar 4-orbital tight-binding model, which is sketched in Fig. 8.1.

**Fig. 8.1** Sketch of a water molecule in the $xz$ plane. The oxygen atom has a $2p_x$ and a $2p_z$ orbital, the hydrogen atoms $H_1$ and $H_2$ have a $1s$ orbital

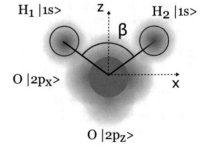

We consider a water molecule lying in the $xz$ plane and include a $1s$ orbital on each hydrogen atom $H_{1,2}$ and $2p_{x,z}$ orbitals on the oxygen atom $O$, with onsite energies $E_{H_{1,2}} = -13.61$ eV and $E_{O_{p_{x,z}}} = -14.13$ eV [16]. We place the Fermi energy $E_F$ halfway between the hydrogen and oxygen onsite energies and from now on we will use it as the zero of energy. The molecular Hamiltonian reads

$$\hat{H}_{H_2O} = \begin{bmatrix} E_{H_1} - E_F & -W_1\cos\theta & W_1\sin\theta & 0 \\ -W_1\cos\theta & E_{O_{px}} - E_F & 0 & W_2\cos\theta \\ W_1\sin\theta & 0 & E_{O_{pz}} - E_F & W_2\sin\theta \\ 0 & W_2\cos\theta & W_2\sin\theta & E_{H_2} - E_F \end{bmatrix}$$

where $W_{1,2} = 1.84\dfrac{\hbar^2}{4m_e R_{OH_{1,2}}^2}$ [16] and $m_e$ is the electron rest mass. $\beta = 104.45°$ [17] is the H-O-H equilibrium angle and $\theta = \dfrac{180°-\beta}{2}$. $R_{OH_{1,2}} = 0.9584$ Å, therefore $W_{1,2} = 3.82$ eV. The resulting molecular orbitals have energies $\left( -4.27\ -3.32\ 3.32\ 4.27 \right)$ eV.

A water molecule presents three normal modes [17]: a low frequency bending mode where $\beta$ oscillates and two higher frequency stretching modes, where the $OH_1$ and $OH_2$ bonds vibrate symmetrically ($\nu = s$) and antisymmetrically ($\nu = a$), as in Fig. 8.2. We include in the simulations only the stretching modes, with frequencies $\hbar\omega_s = 0.4534$ eV and $\hbar\omega_a = 0.4657$ eV and reduced mass is $M_{s,a} = 0.948$ amu [18].

The electron-phonon coupling matrices are

$$\hat{F}_{s,a} = \begin{bmatrix} 0 & \mp\dfrac{F\cos\theta}{\sqrt{2}} & \pm\dfrac{F\sin\theta}{\sqrt{2}} & 0 \\ \mp\dfrac{F\cos\theta}{\sqrt{2}} & 0 & 0 & \dfrac{F\cos\theta}{\sqrt{2}} \\ \pm\dfrac{F\sin\theta}{\sqrt{2}} & 0 & 0 & \dfrac{F\sin\theta}{\sqrt{2}} \\ 0 & \dfrac{F\cos\theta}{\sqrt{2}} & \dfrac{F\sin\theta}{\sqrt{2}} & 0 \end{bmatrix}$$

in which the upper and lower signs indicate respectively the $s$ and the $a$ mode and $F = 1.84\dfrac{\hbar^2}{2m_e R_{OH_{1,2}}^3} = 7.96$ eV/Å.

To get a semiclassical impression of the effect of a phonon on the eigenvalue spectrum, we add $-\hat{F}_{s,a}X_{s,a}$ to the water molecule Hamiltonian $\hat{H}_{H_2O}$, where $X_{s,a}$ is a classical coordinate representing the static phonon displacement. A scan in $X_{s,a}$ samples the elastic effect of a frozen phonon, as shown in Fig. 8.3. In the $s$ mode,

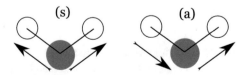

**Fig. 8.2** The water molecule phonon modes included in the simulations, with the arrows pointing towards positive bond displacements $X_{s/a}$. In (s) there is the symmetric mode and in (a) the antisymmetric one

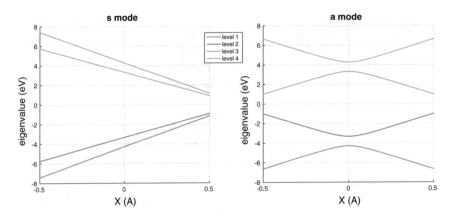

**Fig. 8.3** Water molecule eigenvalues and their elastic variation with the oscillator displacement $X_{s,a}$ in the frozen phonon regime for the $s$ (on the left) and $a$ mode (on the right)

positive displacements bring the hydrogen orbitals further away from the oxygen, reducing the band gap, while the opposite happens for negative $X_s$. In the $a$ mode the bands are symmetric with positive and negative $X_a$. For small $|X_a|$, the eigenvalues are stationary. For increasing $|X_a|$, the badgap reduces while level 1 and 4 get further away.

We can deduce a mean square displacement $\tilde{X}_{v,N} = \sqrt{\langle X^2 \rangle}_N$, for a given $N$. The potential energy for the oscillators in ECEID is always half of the total oscillator energy, therefore $\frac{E_v}{2} = \frac{1}{2}k\langle X_v^2 \rangle_N = (N_v + \frac{1}{2})\frac{\hbar\omega_v}{2}s$. Picking a few values of $N$, in the $s$ mode we have

- $E(N_s = 0) = 0.2267\,\text{eV} \longrightarrow \tilde{X}_{s,0} = 0.070\text{Å}$
- $E(N_s = 2) = 1.133\,\text{eV} \longrightarrow \tilde{X}_{s,2} = 0.156\text{Å}$
- $E(N_s = 5) = 2.494\,\text{eV} \longrightarrow \tilde{X}_{s,5} = 0.231\text{Å}$,

while in the $a$ mode

- $E(N_a = 0) = 0.233\,\text{eV} \longrightarrow \tilde{X}_{a,0} = 0.069\text{Å}$
- $E(N_a = 2) = 1.164\,\text{eV} \longrightarrow \tilde{X}_{a,2} = 0.154\text{Å}$
- $E(N_a = 5) = 2.561\,\text{eV} \longrightarrow \tilde{X}_{a,5} = 0.228\text{Å}$.

### 8.1.2  Embedding Setup

We apply ECEID with OB to a water molecule with the geometry sketched in Fig. 8.4 where we attach $H_1$ to the left lead and $H_2$ to the right lead by $w = 1.0$ eV.

The metallic leads are made of 80 sites with zero onsite energies and hopping $t = -4$ eV, to give an energy bandwidth of 16 eV. The 40 leftmost sites in the left lead and the 40 rightmost ones in the right lead are coupled by $\Gamma = 1.25$ eV to

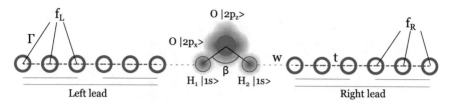

**Fig. 8.4** Schematic of the model system, with a water molecule connected to the left and to the right to metallic leads made of 6 atoms, 3 of which are attached to external probes by $\Gamma$

external probes, following the implementation described in Sect. 5.7. The same $\Gamma$ was introduced in the Hamiltonian driving the auxiliary operators, following Sect. 6.2.

Here we perform ECEID steady state OB calculations, where the leads are populated with zero temperature equilibrium distributions and coupled to probes with ground state distributions $f_{L,R}$ and chemical potentials $\mu_{L,R}$. This setup will be used to generate a bias $V = \mu_L - \mu_R$ across the water molecule, or inject electrons at a specific energy range.

### 8.1.3 Elastic Transmission

We consider the transmission $T_\nu(E, X_\nu)$ from Eq. (B.12) to investigate the purely elastic effect of a phonon frozen at classical displacement $X_\nu$. Here $T_\nu(E, X_\nu)$ represents the probability for an incoming electron with energy $E$ to cross the region with the molecule. It provides an elastic picture, a background on which the inelastic effects can then be overlaid. For $T_\nu(E, X_\nu) = 0$, only inelastic effects allow transmission for an electron.

We perform calculations for the $s$ and the $a$ mode with a maximum displacement $|X_\nu^{max}| = 0.5$ Å, average the transmission over positive and negative displacements $\bar{T}_\nu(E, X_\nu) = (T_\nu(E, X_\nu) + T_\nu(E, -X_\nu))/2$ and show the results in Fig. 8.5.

For $X_\nu = 0$, there are 4 transmission channels corresponding to the water molecule energy levels. The introduction of the phonon modes alters the position of these channels in energy with a marked difference between the two modes. At small $X_\nu$, the transmission peaks of the $s$ mode branch out reducing the bandgap, while in the $a$ mode they don't exhibit a significant change. At large $X_\nu$, the transmission in the $s$ mode presents an increasingly small bandgap with a substantially unchanged transmission peak intensity, whereas the peaks in the $a$ mode decay in intensity with $X_\nu$. Symmetry plays an important role in resonant systems such as this. Symmetry breaking, as in the $a$ mode case, can lead to strong reductions in transmission probability.

Another quantity that offers insights about the elastic properties of the modes is the current, which can be found from integrating the transmission [19]

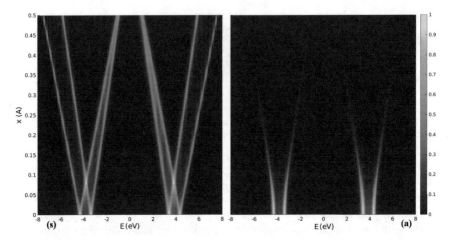

**Fig. 8.5** Frozen phonon calculation of the transmission $\bar{T}_\nu(E, X_\nu)$ as a function of energy, averaged over positive and negative displacement, for the $s$ mode (s) and the $a$ mode (a). Lighter colours correspond to regions of high transmission, whereas darker colours are related to a low transmission

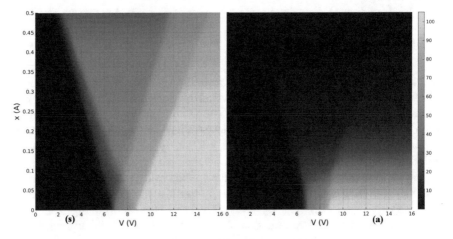

**Fig. 8.6** Elastic current in ($\mu A$) for the $s$ mode (s) and $a$ mode (a)

$$I_\nu^{\text{el}}(V, X_\nu) = \frac{e}{h} \int_{E_F - \frac{eV}{2}}^{E_F + \frac{eV}{2}} dE \; \frac{\bar{T}_\nu(E/2 - E_F, X_\nu) + \bar{T}_\nu(-E/2 - E_F, X_\nu)}{2} \quad (8.1)$$

and is plotted in Fig. 8.6. In the $s$ mode a non-zero current develops at lower and lower $V$ for increasing $X_s$, due to the bandgap getting smaller. The high voltage current presents a slow decreasing plateau with $X_s$, related to the slow decay of the transmission peak. The $a$ mode doesn't show a clear reduction in the minimum $V$ necessary to allow a current and, for high $X_a$, presents a rapidly decaying current. This was expectable after seeing the rapidly decaying transmission peaks in Fig. 8.5a.

It is natural to wonder to what extent the inelastic effects alter the transmission scenario. We introduce inelasticity in the next section and compare elastic IV curves with ECEID ones.

## 8.1.4 ECEID Comparison with Elastic Averages

For comparing the elastic current $I_\nu(V, X_\nu)$ in Eq. (8.1) with its inelastic equivalent from ECEID $I_\nu(V)$, first we need to eliminate the $X_\nu$ dependency in the elastic result by averaging it away. The easiest procedure considers an average over two current values: the positive and the negative value of the mean square displacement $\tilde{X}_{\nu,N}$

$$I_{\nu,N}^{2P}(V) = \frac{I_\nu(V, \tilde{X}_{\nu,N}) + I_\nu(V, -\tilde{X}_{\nu,N})}{2}. \tag{8.2}$$

We call it 2-Point (2P) average. Another possibility is a top hat (TH) distribution function with standard deviation $\tilde{X}_{\nu,N}$, so that the resulting arithmetic average of the current is

$$I_{\nu,N}^{TH}(V) = \frac{1}{2\sqrt{3}\tilde{X}_{\nu,N}} \int_{-\sqrt{3}\tilde{X}_{\nu,N}}^{\sqrt{3}\tilde{X}_{\nu,N}} I_\nu(V, X_\nu) dX_\nu. \tag{8.3}$$

An averaging that takes into account the spatial probability distribution of a phonon with $N_\nu$ uses the quantum harmonic oscillator wavefunction (HOWF)

$$\Psi_{N_\nu}(X_\nu) = \frac{1}{\sqrt{2^{N_\nu} N_\nu!}} \left(\frac{M_\nu \omega_\nu}{\pi \hbar}\right)^{1/4} e^{-M_\nu \omega_\nu X_\nu^2/(2\hbar)} H_{N_\nu}\left(\sqrt{\frac{M_\nu \omega_\nu}{\hbar}} X_\nu\right) \tag{8.4}$$

where $H_{N_\nu}$ is an Hermite polynomial. It weighs the elastic current as

$$I_{\nu,N}^{HOWF}(V) = \frac{\int_{-X_\nu^{max}}^{X_\nu^{max}} dX_\nu |\Psi_{N_\nu}(X_\nu)|^2 I_\nu^{el}(V, X_\nu)}{\int_{-X_\nu^{max}}^{X_\nu^{max}} dX_\nu |\Psi_{N_\nu}(X_\nu)|^2}. \tag{8.5}$$

where the cutoff $X_\nu^{max}$ is large enough to have $|\Psi_{N_\nu}(X_\nu^{max})|^2 \approx 0$. This average is the most physical of the three as it takes into account the oscillator wavefunction explicitly. In Fig. 8.7 we show the average curves for the $a$ mode and $N = 0, 2, 5$.

ECEID OB simulations with a bias $V$ establish a steady state current at long times. Scanning over $V$, it is possible to determine $I$ for different phonon configurations with $\dot{N} = 0$ and fixed $N = 0, 2, 5$. We compare them with elastic averages results in Fig. 8.8. In ECEID, $I(V, N)$ is determined both by the elastic background and by the inelastic interaction of the phonon with the incoming stream of electrons. When the elastic results agree with ECEID, the elastic contribution must be dominant. When they differ, ECEID could display relevant inelastic effects but there's no guarantee

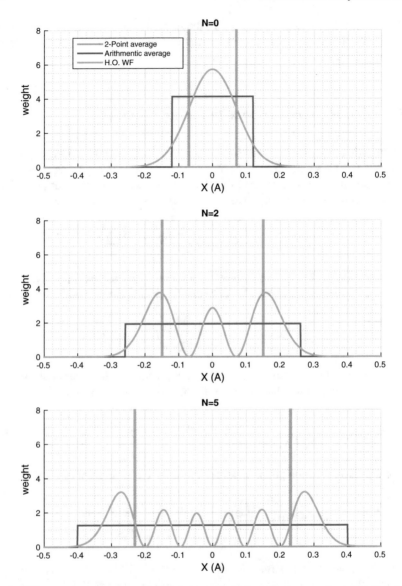

**Fig. 8.7** Curves to average the $X_\nu$ dependency in $I_\nu(V, X_\nu)$ for the $a$ mode at $N = 0, 2, 5$. In orange the 2PA, in red the TH and in green the HOWF

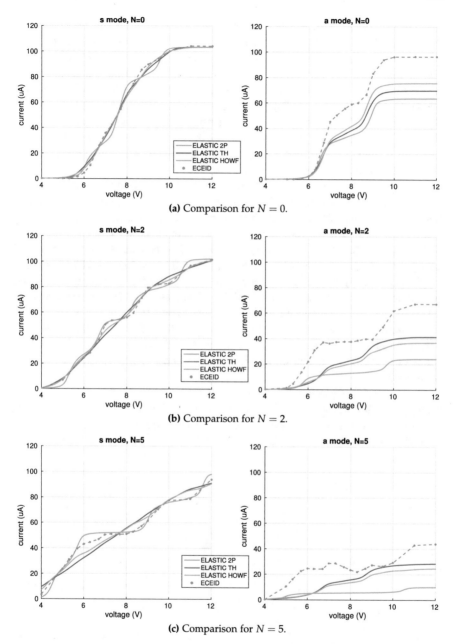

**(a)** Comparison for $N = 0$.

**(b)** Comparison for $N = 2$.

**(c)** Comparison for $N = 5$.

**Fig. 8.8** ECEID IV curves (blue) compared to elastic averages 2P (orange), TH (red) and HOWF (green). It is remarkable that in some cases the ECEID current exceeds the elastic one. This inelastic effect on transmission will be analysed in Sect 8.2

that inelasticity is the main cause for divergence. To rule out any ECEID wrong calibration, Sect. 8.1.5 offers a comparison of ECEID with two known limits.

The *s* mode at $N = 0$ displays a remarkable agreement between ECEID and both HOWF and TH averages. The $2P$ case has a few different features that can be due to its sampling of only two different $X_s$. In the *a* mode at $N = 0$, even if the current curves show similarities in shape, the ECEID current is consistently larger than the elastic ones. This hints to the *a* mode featuring a strong inelastic contribution. From the transmission results we noticed that the elastic contribution in the *a* mode disappears quickly for increasing $X_a$. Results for higher $N$ sample higher $X_a$ and help in further isolating the inelastic effects.

At $N = 2$, in the *s* mode the ECEID curve is still very close to the elastic averages, whereas it shows large differences in the *a* mode. In the latter case both the ECEID and the elastic averages display a clear drop in current, consistent with the transmission result, with the ECEID current being still larger than the elastic cases.

For $N = 5$, the *s* mode starts to show a slight disagreement between ECEID and the elastic curves, with the HOWF case displaying the best agreement. The *a* mode exhibits again a large decrease in current, with the ECEID result featuring details that are not present in the elastic curves, including the presence of a negative differential resistance region (an increasing $V$ leads to a decreasing $I$).

To summarize these results, in Fig. 8.9 we join the HOWF and the ECEID IV curves for different $N$. There we get a visual confirmation of how well do the ECEID simulations reproduce the elastic results in the *s* mode and how different they are in the *a* mode.

To extract the inelastic contribution from the data, we analyze the percentage change in current from the HOWF average to ECEID in Fig. 8.10. In the *s* mode, excluding data at low $V$ where the very low currents create large percentage changes, the inelastic difference between ECEID and HOWF is low and mostly within a 10%

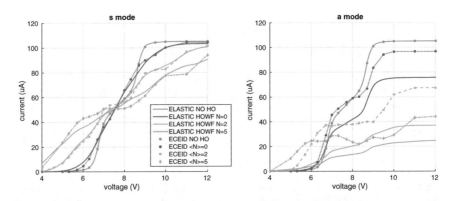

**Fig. 8.9** Comparison of ECEID $IV$ curves with elastic HOWF averages for the *s* mode (left) and the *a* mode (right). Solid lines indicate elastic calculations, dots (and connecting dashed lines) correspond to ECEID. The case without a phonon is in blue, $N = 0$ in red, $N = 2$ in green and $N = 5$ in yellow

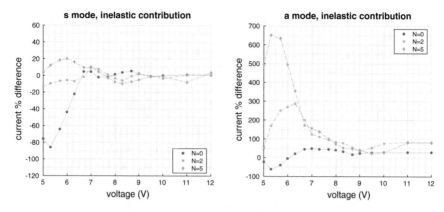

**Fig. 8.10** Percentage difference between the current in ECEID and in the elastic HOWF for $N = 0$ (red), $N = 2$ (green), and $N = 0$ (yellow). The $s$ mode is on the left and the $a$ mode on the right

range. The $a$ mode shows a much more intense inelastic contribution, often close to a 100% increase.

The two modes belong to two different conduction regimes. The $s$ mode presents a quite high transmission for all oscillator displacements, so the effect of the phonon tends to lower the current, especially at high $N$ where the inelastic scattering increases. On the other hand, the $a$ mode presents a lower elastic transmission in the presence of elastic phonon displacements, therefore the inelastic effect of a phonon tends to enhance the current. In the two regimes the physics is different and the introduction of inelastic scattering affects the conductance in opposite ways, a well known result that was observed, for example, in a gold nanowire [20, 21]. In the low transmission limit an inelastic channel increases the conductance, while in the high transmission limit it enhances backscattering and reduces the conductance.

## 8.1.5 The Landauer and the High Mass Limit

In general, ECEID has no guarantee of agreement with the elastic results. ECEID is expected to agree with the elastic case only when the inelastic contribution goes to zero. Here we test two limits where agreement is expected.

First, in the case without phonons, we expect ECEID with OB to recover the Landauer result for a water molecule.[1] In Fig. 8.9 the blue dots represents the ECEID current without a phonon at different $V$ and the solid blue curve is the Landauer elastic current. The curves are indeed superimposable. This comparison provides a physical validation for the setup and confirms that the length of the leads and $\Gamma$ are a good

---

[1]Here, by Landauer result we mean the elastic current as an integral of the transmission from Eq. (8.1) without any phonon.

choice. This limit can be seen as a starting point over which inelastic effects are juxtaposed.

The other limit that we probe is the High Mass Limit (HML). By increasing the phonon's mass and decreasing its frequency, while keeping the phonon's energy and its mean square displacement constant, we expect the phonon to become more and more static, until it converges to the behaviour of an immovable object that scatters incoming electrons elastically. We set $\tilde{X}_s = 0.156$ Å and $\tilde{X}_a = 0.154$ Å and probe different masses starting from the unchanged one $M_{s,a} = 0.948$ (M = 1x case) up to one thousand times heavier (M = 1000x case) and show the IV curves in Fig. 8.11. The $s$ mode converges very fast: the M = 10x case looks to be already well converged.

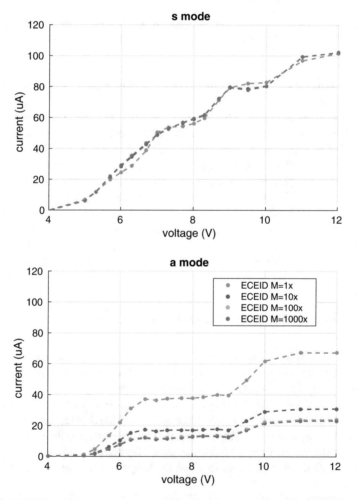

**Fig. 8.11** ECEID High Mass Limit IV plots for the $s$ mode and the $a$ mode. The usual mass M = 1x is in blue, the M = 10x case in red, the M = 100x in green and the M = 1000x in purple

The $a$ has a slower convergence and a bigger result variation for an increasing mass; its M = 100x case looks to be at convergence.

To link these results to the elastic average curves, we notice that in [22] it was analytically found that, in the HML, CEID converged to the elastic 2P average. We check that ECEID does the same by going through a similar derivation.

In the HML, the DM still obeys its Liouville EOM (5.5) which remains unchanged and $\dot{N}_\nu$ follows Eq. (5.10). The limit $M_\nu \to 0$ and $\omega_\nu \to \infty$ with $M_\nu \omega_\nu^2 \langle X_\nu \rangle^2$ constant makes $M_\nu \omega_\nu \to \infty$ so that $\dot{N}_\nu \to 0$. Therefore $N_\nu(t) = \bar{N}_\nu$ is a constant in time, with $\bar{N}_\nu = M_\nu \omega_\nu \langle X_\nu \rangle^2 / \hbar \to \infty$. The remaining quantity that has to be tracked in this limit is $\hat{\mu}_\nu$ from Eq. (5.34). $\hat{A}_\nu^s(t)$ is finite and, as it appears in $\hat{\mu}_\nu$ with a $1/(M_\nu \omega_\nu) \to 0$ prefactor, it does not contribute in the HML. In this limit the we then have

$$\hat{\mu}_\nu(t) \approx \frac{i}{M_\nu \omega_\nu} \hat{C}_\nu^c(t) \approx \frac{i \bar{N}_\nu}{M_\nu \omega_\nu} f(t) = \frac{i \langle X_\nu \rangle^2}{\hbar} f(t) \qquad (8.6)$$

where $f(t) = \int_0^t [\hat{F}_\nu^{\tau-t}, \hat{\rho}_e^{\tau-t}] \, d\tau$.

In the semiclassical problem with phonon displacement $X_\nu$, we have

$$\hat{\mu}_\nu(t) = \int X_\nu \hat{\rho}(X_\nu, t) \chi(X_\nu) dX_\nu \qquad (8.7)$$

where $\hat{\rho}(X_\nu, t)$ is the full DM and $\chi(X_\nu)$ is the distribution of phonon displacements. We pick $\chi(X_\nu) = \frac{1}{2}(\delta(X_\nu + \langle X_\nu \rangle) + \delta(X_\nu - \langle X_\nu \rangle))$, which corresponds to the 2P average. By plugging the exact form of the DM from Eqs. (5.4) into (8.7) and integrating over $X_\nu$, we obtain

$$\hat{\mu}_\nu(t) = \frac{i \langle X_\nu \rangle^2}{\hbar} \int_0^t [\hat{F}_\nu^{\tau-t}, \hat{\rho}_e^{\tau-t}] \, d\tau \qquad (8.8)$$

which agrees with the expression (8.6). In the HML, ECEID converges to the semiclassical case with a 2P average for the phonon.

In Fig. 8.12, we compare 2P average IV curves with ECEID in the HML. The $a$ mode converges with the elastic result at about M = 100x. The $s$ mode at M = 1000x is close to the elastic case, but still not superimposable. In the HML we expect exact convergence, and, to check if the discrepancy could be due to the OB setup, we increase the length of the leads to 250. The longer leads curve gets much closer to the elastic case, indicating that, in the infinite lead limit, there is convergence. Therefore, ECEID simulations verify the HML.

The fact that ECEID recovers the elastic limit is a strong verification of the method. This convergence to an exact result is not obvious given the approximate nature of ECEID and the extent of its approximations. The limits in this section provide a solid stepping stone for the following inelastic calculations.

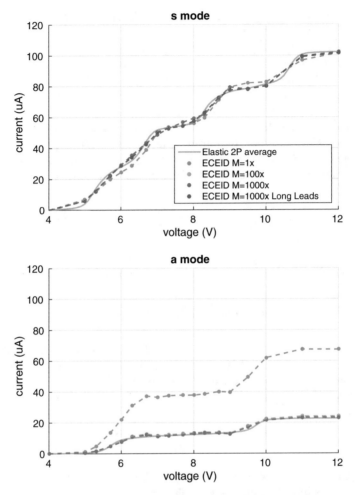

**Fig. 8.12** Comparison of ECEID in the HML with elastic 2P average IV curves for the *s* mode and the *a* mode. The curve in green is the elastic one, the blue one is the M = 1x case, the yellow one the M = 100x, the purple one the M = 1000x and the red one the M = 1000x with longer left and right leads of 250 atoms

### 8.1.6   Current Assisted Phonon Heating

We apply a bias to the molecule in the undamped limit by not freezing $\dot{N}$ at zero and starting from $N(0) = 0$. The phonon modes can now interact with the injected electrons, until the system reaches a steady state.

In Fig. 8.13a, we compare the steady state current for the two modes with the elastic Landauer result without phonons. At low voltages, both ECEID simulations show a slight current increase over the elastic case, indicating an inelastic mechanism

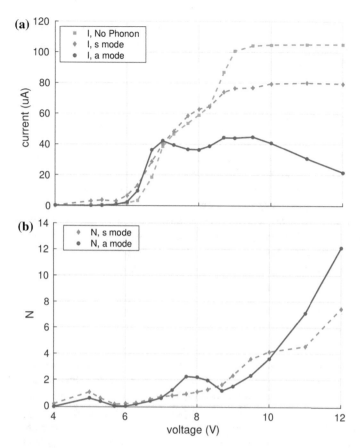

**Fig. 8.13** ECEID water molecule simulations with a bias $V$, the vibrational modes starting from $N(0) = 0$ and $\dot{N}$ free to vary. In blue the $s$ mode, in red the $a$ mode and in yellow the elastic Landauer case (without phonons). In **a** we show the steady state current and in **b** the steady state $N$

that helps the electrons flow. At about $V = 7$ V, the current in the $a$ mode starts to decrease for increasing biases. This phenomenon is known as negative differential resistance [22, 23].

For high biases, the steady state current in both modes either decreases or reaches a plateau that is less than the elastic value. The presence of the phonons enhances dissipation. By looking at the corresponding steady state value of $N$ in Fig. 8.13b, we see that, for high biases, $N$ increases greatly, especially for the $a$ mode. Moreover, the equilibration time for high biases increases too.

This dramatic heating accompanied by enhanced relaxation times was also observed in CEID simulations on a resonant system [22] and led the authors to conclude that at high enough biases a strong local heating in the system can occur. In realistic systems such a heating is instrumental in the formation of structural damage. For example, in the field of radiation damage to biological systems, the heating of the environment caused by secondary electrons can cause cellular damage. This motivates us in investigating larger systems including more than one water molecule, such as a water chain.

## 8.2   Water Chain

We used DFT simulations to determine the geometry of a water chain and, at first, we implemented the same tight-binding model as the one for the water molecule. The results with this model are in Appendix C. The tight-binding electronic structure did not show a good agreement with the corresponding band structure from DFT simulations, so we decided to improve the model. In the following section, we employ a more advanced O-H hopping and obtain a tight-binding band structure in better agreement with DFT. We proceed by injecting electrons in the chain studying their elastic and inelastic interaction with the water chain.

### 8.2.1   The Model

We connect water molecules to form a planar chain, whose equilibrium geometry was determined with DFT simulations[2] and is shown in Fig. 8.14. The resulting electronic structure from DFT features a HOMO-LUMO gap of 6 eV followed by a 1.6 eV-wide group of unoccupied states. These are separated from the continuum by a 3 eV gap, which probably originates from the 1-dimensional geometry of the chain. We set as our aim to reproduce the presence of the isolated lowest unoccupied band in our tight-binding model and to use it for the electron injection in the examples.

The orbital onsite energies and the $E_F$ are exactly as in the water molecule case, while the hoppings employ a more advanced model. The geometry parameters such as the bond lengths and the H-O-H angle here come from the DFT simulations. The Hamiltonian for the $j$th water molecule in the chain reads

$$
\hat{H}_j = \begin{bmatrix}
E_{H_1} - E_F & \mp U_1 \cos\theta & U_1 \sin\theta & 0 \\
\mp U_1 \cos\theta & E_{O_{px}} - E_F & 0 & \pm U_2 \cos\theta \\
U_1 \sin\theta & 0 & E_{O_{pz}} - E_F & U_2 \sin\theta \\
0 & \pm U_2 \cos\theta & U_2 \sin\theta & E_{H_2} - E_F
\end{bmatrix}
$$

where upper signs match an odd $j$ and lower signs an even $j$, $U_{1,2}$ are the hoppings between the oxygen and the hydrogen atoms $H_{1,2}$, $\beta = 105.8°$ is the H-O-H angle and $\theta = (\pi - \beta)/2$. $H_1$ points out of the chain with a bond length $R_{OH_1} = 0.97$ Å and $H_2$ forms the chain's backbone with $R_{OH_2} = 1.00$ Å. We choose a O-H hopping

---

[2]We used the software CP2K [24] with PBE functional and a 6311G** basis.

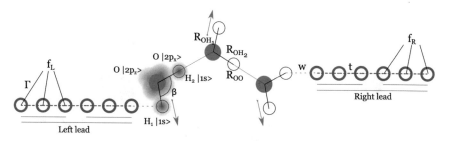

**Fig. 8.14** Reduced schematic of our model system, with a water chain of 3 molecules connected to its left and its right to metallic leads made of 6 atoms, 3 of which attached to external probes. See text for the dimensions of the system in the actual simulations. Each oxygen atom has a $2p_x$ and a $2p_z$ orbital, while the hydrogen atoms $H_1$ (pointing out of the chain) and $H_2$ (pointing in the chain) have a $1s$ orbital. The red arrows depict the phonon mode included in the calculations, with the $R_{OH_1}$ bonds vibrating in phase [9]

$$U(R) = V \left(\frac{R_0}{R}\right)^2 \exp\left(2\left(-\left(\frac{R}{R_C}\right)^4 + \left(\frac{R_0}{R_C}\right)^4\right)\right) \tag{8.9}$$

where $V = 1.84\,\hbar^2/(2m_e R_0^2)$, $m_e$ is the electron mass, $R_0 = R_{OH_1}$ is the equilibrium bond length and the critical length is set to $R_C = 1.8\,R_{OH_1}$ [16, 25]. This makes $U_1 = U(R_{OH_1}) = 7.45$ eV and $U_2 = U(R_{OH_2}) = 6.84$ eV.

The distance between the oxygen atoms is $R_{OO} = 2.67$ Å, so the inter-molecular hopping between $O2p_z$ and $H_2$ is $U(R_{OO} - R_{OH_2}) = 0.57$ eV, while the inter-molecular hopping between the $O2p_x$ orbital and $H_2$ is zero by construction. This form of hopping has the advantage of recovering the result from Harrison's solid state table [26] $U(R_0) = V$ for equilibrium bond lengths, while decaying exponentially with large distances, suppressing intermolecular O-H interactions.

A DFT vibrational structure calculation for a 10 molecule water chain with periodic boundary conditions produces a set of phonon bands. One of them consists of phonon modes that predominantly involve out of chain hydrogens $H_1$ moving in and out of the chain. The modes in that band present closely clustered frequencies lying within a fraction of a percent of each other. For our present calculations, we are going to consider a fictitious mode (since the actual modes and their phases will of course depend on the specific boundary conditions) in which all $H_1$ are vibrating in phase with each other, as indicated in Fig. 8.14. We considered different choices of relative phases for the vibrating $H$ atoms, but these did not produce a qualitative change in the results. We take a representative value for the frequency $\hbar\omega = 0.473$ eV and for the vibrational reduced mass $M = 1$ amu, for that band of modes. A block of $\hat{F}$ for the $j$th water molecule in the chain reads

$$\hat{F}_j = F \begin{bmatrix} 0 & \mp\cos\theta & \sin\theta & 0 \\ \mp\cos\theta & 0 & 0 & 0 \\ \sin\theta & 0 & 0 & 0 \\ 0 & 0 & 0 & 0 \end{bmatrix}$$

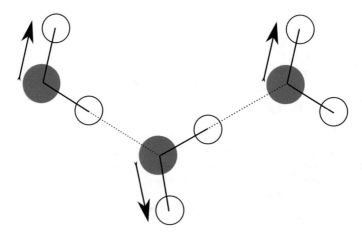

**Fig. 8.15** The high frequency phononic mode for a 3-molecule water chain with the $R_{OH_1}$ bonds vibrating in phase

where $F = C \frac{\partial U(R)}{\partial R}\Big|_{R=R_{OH_1}} = 6.60\,\text{eV/Å}$, the upper signs correspond to an odd $j$, the lower ones to an even $j$ and the factor $C = 1/\sqrt{10}$ arises because our phonon involves the motion of all the 10 water molecules in the chain. The zero point amplitude of the phonon is $\sqrt{\langle X^2 \rangle}_{N=0} = 0.064$ Å, while its root mean square displacement at $N = 2$ is $\sqrt{\langle X^2 \rangle}_{N=2} = 0.144$ Å.

With the above value of $F$, in some simulations we observed violent oscillations in the current that prevented the formation of a steady state in the system. We believe that these oscillations are due to a limitation of the method approximations in treating coherences for large $F$. We can determine a critical $F$ by using a qualitative condition that compares a typical electron-phonon transition rate with the inverse oscillator period $\frac{2\pi}{\hbar} F_c^2 X^2 D \simeq \frac{\omega}{2\pi}$. If we pick as a typical phonon displacement $X \simeq \sqrt{\langle X^2 \rangle}_{N=0}$ and a density of states $D \simeq \frac{1}{B}$ with $B \simeq 16$ eV the total bandwidth of the water chain, we obtain $F_c \simeq 7$ eV/Å. System dynamics with values about and above $F_c$ would involve higher-order processes and coherences that ECEID cannot handle. The value above is indeed close to $F_c$. As a cure, consistent with the approximations in the method, we halve $F$ to 3.30 eV/Å. The electron-phonon transition rates scale, to lowest order, quadratically with $F$; therefore, the above change would reduce the inelastic transition rates by a factor of 4. Nevertheless, the underlying physics of the phenomenon that we are interested in, namely the inelastic electron injection in elastically forbidden energy ranges, remains unchanged.

## 8.2.2 Simulation Details

We work with a chain made of 10 water molecules. Its electronic eigenvalues form 4 bands that are symmetric with respect to $E_F$. The 2 upper bands, the first conduction band (FCB) and the second conduction band (SCB), are shown in Fig. 8.16. The water chain eigenstates are initially populated at their ground state with a half filled band. DFT calculations of the excited states of a corresponding system show the formation of a band analogous to the FCB and, above it, a continuum of states which is not captured in our simple TB model because of the reduced number of basis functions for water empty states.

Our TB model of the chain can reproduce the feature of the FCB and is in qualitative agreement with the band structure from DFT. The band gap in the TB chain is about 11 eV, which compares well with the accepted experimental value of 7 eV for liquid water and 9 eV for ice [27]. In our DFT simulations the band gap is 6 eV. The FCB bandwidth is 0.8 eV in TB, against a value of 1.6 eV from DFT. The gap above the FCB is 1.6 eV in the TB model and 3 eV in DFT.

At higher energies the band structure of TB differs significantly from the DFT one, because of the minimal basis set used that comprises only $s$ orbitals for the H and $p_x$ and $p_z$ for the O. The continuum in DFT arises from additional $s$ and $p$ as well as $d$ orbitals in the basis set. This section focuses on injecting electrons in the FCB and evaluating the inelastic contribution of the phonon. Effects involving the artificial SCB will be commented on later.

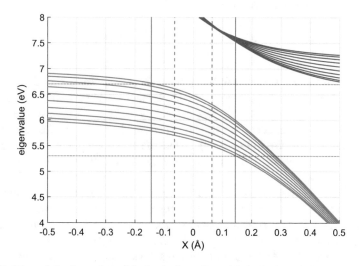

**Fig. 8.16** Water chain eigenvalues of the levels above $E_F$ and their elastic variation with the phonon mode generalized coordinate $X$, with the FCB shown at the bottom of the figure and the SCB on top. The dashed vertical lines indicate the phononic zero point motion $\sqrt{\langle X^2 \rangle}_{N=0}$ and the solid ones $\sqrt{\langle X^2 \rangle}_{N=2}$. The horizontal dotted lines indicate the energies $E = 5.3\,\text{eV}$ and $E = 6.7\,\text{eV}$ that will be used in the electron-gun injection section [9]

To examine the effect of a classical phonon on the eigenspectrum, we add $-\hat{F}X$ to the water chain Hamiltonian, where $X$ is a classical generalized coordinate. A scan in $X$ samples the elastic effect of a frozen phonon. In Fig. 8.16 we show the variation with $X$ of the eigenvalues in the FCB. A positive $X$ makes $R_{OH_1}$ extend, reducing the FCB bandwidth and narrowing the gap with the valence band. A negative $X$ makes the FCB shift slightly upwards in energy.

Next, we connect the water chain to metal leads, as sketched in Fig. 8.14, so that $H_1$ on the left and $H_2$ on the right of the chain are linked to the neighbouring metal site by $w = -2$ eV. The metallic leads serve as a tool for injection and, initially, they are not populated. They are made of 80 sites each, with a single state per site and a nearest neighbour hopping $t = -4$ eV. Their onsite energy is 10 eV, so that the metallic energy band ranges from 2 to 18 eV, overlapping only with the water's conduction bands. The 40 leftmost sites in the left lead and the 40 rightmost ones in the right lead are coupled by $\Gamma = 1.25$ eV to external probes for the OB.

### 8.2.3   Electron-Pulse Injection

We begin with probably the most intuitive picture of electron injection, namely firing actual electronic wavepackets from the left lead towards the water chain and propagating the ECEID EOM. Initially, we introduce in the empty leads an electron pulse in the form of a spin-degenerate wavepacket moving from the left lead towards the water chain. The pulse has the following form

$$|\Psi_0\rangle = c \sum_n e^{-(n-n_0)^2/2\sigma^2} e^{ikn} |n\rangle \qquad (8.10)$$

where $n$ spans the left lead atomic basis, $n_0 = 55$ is the pulse's central site, $\sigma = 12$ sites, $k \in [0, \pi]$ is a dimensionless crystal momentum and $c$ is the normalization factor. By scanning over $k$, we can vary the energy of the incident wavepacket which has an initial full width at half maximum in energy of about 1 eV. We keep only the extraction term in the OB in Eq. (5.49), so that the backscattered or transmitted parts of the wavepacket reaching the ends of the leads can get absorbed and not reflected back into the water.

To measure electron absorption into the water, we sum over the occupations of the FCB. We call this quantity excess electron population (EEP) and show it at different times in Fig. 8.17a. We perform a phonon-free elastic simulation (dotted line) and inelastic simulations where the phonon is initialized at $N(0) = 0$ (dashed line) or $N(0) = 2$ (solid line). At $t = 5$ fs the EEP displays a peak for an incident energy in the middle of the FCB. It also displays electron absorption for incident energies well outside the FCB. The inelastic contribution to injection in this elastically forbidden range is intertwined with elastic effects due to the energy width of the pulse. In fact, the elastic curves are not significantly different from the inelastic ones at this level of

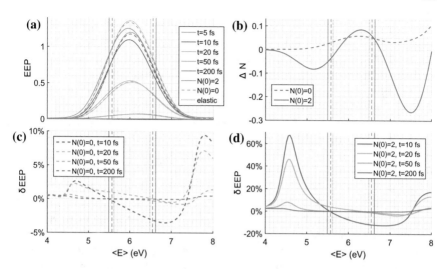

**Fig. 8.17** Electron-pulse simulations where Gaussian wavepackets are injected into the water chain. The vertical dotted lines mark the edges of FCB for $X = 0$; the dashed vertical lines mark the range spanned by FCB for $|X| < \sqrt{\langle X^2 \rangle}_{N=0}$; the solid vertical lines mark this range for $|X| < \sqrt{\langle X^2 \rangle}_{N=2}$. Panel (**a**): FCB excess electron population for wavepackets with an average energy $\langle E \rangle$ at a range of times. Solid lines correspond to the phonon starting at $N(0) = 2$, dashed lines to $N(0) = 0$ and dotted lines to the elastic case. Panel (**b**): phonon occupancy variation $\Delta N$ at $t = 200$ fs for initial $N(0) = 0$ (dashed) and $N(0) = 2$ (solid). Panel (**c**): difference between the inelastic excess electron population with $N(0) = 0$ and the elastic one at different times, normalized by the elastic case at $t = 5$ fs. Panel(**d**): same as panel (**c**), with the inelastic case of $N(0) = 2$ [9]

analysis. With time progressing, the EEP decreases in magnitude as electrons leak back out into the leads, until, at $t = 200$ fs, the EEP drops close to zero in all cases.

Nevertheless, the electron pulse leaves a long lasting inelastic mark on the phonon occupancy variation $\Delta N(t) = N(t) - N(0)$, shown at $t = 200$ fs in Fig. 8.17b. $\Delta N(t)$ displays a radically different behaviour depending on the initial condition $N(0)$ and on the pulse average energy $\langle E \rangle$. Electron injection in the elastically forbidden range, outside the FCB, hinges on the phonon, as the electron-phonon interaction enables electrons to access the water states via phonon emission or absorption. This inelastically-assisted electron injection furthermore is controlled by the effective vibrational temperature, as the value of $N(0)$ determines which phonon-assisted processes are allowed.

For pulses below the water FCB edge, $\langle E \rangle < 5.6$ eV, an electron must absorb phonons to enter the water; therefore the inelastic hopping can only be activated if $N(0) > 0$. Indeed, we see that the dashed curve representing $N(0) = 0$ in Fig. 8.17b remains close to zero at low energies and starts increasing for pulses with $\langle E \rangle \simeq 5$ eV. At that energy range, the high energy components of the pulse can enter the FCB elastically and trigger further inelastic processes. On the other hand, the solid curve, for $N(0) = 2$, shows a clear dip in $\Delta N$ at incident energies about $\hbar \omega$ below FCB, highlighting electron injection assisted by phonon absorption.

For incident pulses above the upper band edge, $\langle E \rangle > 6.5$ eV, electrons have to emit a phonon to enter the water FCB, so one expects a $\Delta N > 0$. However, the higher energy SCB (see Fig. 8.16) obscures the phonon emission process in this range by mixing it with phonon absorption, leading to injection into SCB.

To isolate the inelastic contribution in the EEP, we introduce $\delta$EEP: the difference between the inelastic EEP and the elastic EEP, normalized by the elastic EEP at $t = 5$ fs. This quantity measures the effective inelastic contribution to injection. In Fig. 8.17c–d, we show $\delta$EEP for $N(0) = 0$ and $N(0) = 2$ respectively. The latter shows an intense peak at $\langle E \rangle \simeq 4.5$ eV, corresponding to phonon absorption by the high-energy components of the incident wavepacket. This signature peak is much weaker in the $N(0) = 0$ case. Thus $N$, and the effective phonon temperature, is a key controlling factor for the intensity of phonon-assisted injection. Above the FCB, we see another increase in $\delta$EEP marking inelastic injection in the FCB, but mixed with the concomitant injection in the SCB.

The decay time of the injected electrons shows a dependency on $\langle E \rangle$, with $\delta$EEP for pulse energies outside the FCB decaying slower than for energies inside the FCB. This behaviour suggests an energy dependent lifetime of the FCB states, which we will explore more thoroughly later.

In the electron-pulse simulations, the energy spread of the wavepackets obscures the inelastic effects on the FCB by mixing them with elastic contributions and the injection in the SCB. To overcome these limitations, and probe the inelastic injection mechanism further, we now exploit the OB to send in a steady incident electron beam with a sharp energy spectrum.

### 8.2.4   Electron-Gun Injection

We make use of the electronic OB Eq. (5.49) to inject a steady electron beam, as opposed to the single wavepacket above. Here the leads are coupled to baths kept at zero electronic temperature and zero electrochemical potential. Thus $f_{L/R}(E)$ are the corresponding equilibrium electronic distributions. However, in addition, $f_L(E)$ contains a top-hat spike between $E = \epsilon - \delta\epsilon$ and $E = \epsilon + \delta\epsilon$. This setup provides a steady electron flux hitting the water chain. We concentrate on the interesting scenario found earlier when the incoming electron energies are just above or below the water FCB, so that phonon-activated electron injection is the dominant process. We aim the electron gun at $\epsilon = 5.3$ eV with $\delta\epsilon = 0.1$ eV, just below the chain FCB $5.6 < E < 6.5$ eV and just above it at $\epsilon = 6.7$ eV, as can be seen from the horizontal dotted lines in Fig. 8.16.

In Fig. 8.18a–b we show the current measured on the right and on the left of the water chain in ECEID simulations, with the initial phonon occupancies $N(0) = 0$, $N(0) = 2$ considered earlier, together with a purely elastic, phonon-free simulation. In all cases, the injection energy is out of the chain's elastic transmission range, so the elastic current remains very close to zero as expected. At the lower injection energy $\epsilon = 5.3$ eV, if $N(0) > 0$, electrons can hop into the water FCB through

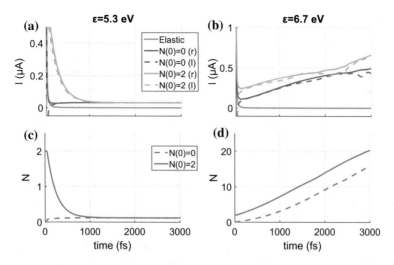

**Fig. 8.18** Electron-gun simulations where electrons are injected within a narrow energy window, just above or below the water FCB. Panels (**a**) and (**b**): current measured in the second metal bond on the right (solid) and on the left (dashed) of the chain with the electron gun aimed at $E = 5.3 \pm 0.1$ eV and $E = 6.7 \pm 0.1$ eV, respectively. Elastic ECEID simulations are in blue, $N(0) = 0$ in red and $N(0) = 2$ in green. Panels (**c**) and (**d**): dynamics of $N$ corresponding to (**a**) and (**b**), with $N(0) = 0$ (dashed) and $N(0) = 2$ (solid) [9]

phonon absorption. Indeed, at $N(0) = 2$, the current decreases in time, with the phonon cooling down, as shown in Fig. 8.18c. This current-assisted cooling may look counterintuitive if one imagines a current to always heat its surroundings, but in fact current-assisted cooling in molecular systems is not unfamiliar [28]. Here the phonon cooling and the nonzero current are a signature of the phonon absorption injection mechanism.

At $N(0) = 0$ and $\epsilon = 5.3$ eV, even if no injection is expected, there is a small increase in current accompanied by an increase in $N$, until both the current and $N$ eventually reach a steady state matching the one for $N(0) = 2$. This is caused by the broadening of the FCB due to the presence of the leads, together with the fact that the electron-gun window of populated states in the leads effectively constitutes an electronic excitation above the ground state of the system. Therefore, thermodynamically, a tendency of the phonon to equilibrate at a raised energy above the vibrational ground state is to be expected. The broadening of the FCB density of states—and the small heating effect starting from $N(0) = 0$—decrease with decreasing water-metal coupling ($w$).

In the $\epsilon = 5.3$ eV results, the currents evaluated on the left and on the right of the chain are superimposable: the system dynamics can be approximated by a series of steady states. This does not happen for $\epsilon = 6.7$ eV, where left and right currents tend to slightly differ at all times, indicating that the system dynamics does not manage to remain in an electronic steady state during the heating of the phonon. The mechanism at play now is electron injection into water through phonon emission, with a positive feedback mechanism. The incoming electron flux causes phonon heating through emission, and a resultant current due to the electrons inelastically injected in the water. The increase in $N$, in turn, increases the electron-phonon scattering rate (from the Golden Rule), further increasing the inelastic electron current. In the present model, the water is not coupled to a vibrational thermal bath and this leads to the dramatic phonon heating see in Fig. 8.18d.

Of course, the vibrational energy cannot increase indefinitely, but the principle remains: a potentially powerful injection mechanism from elastically forbidden energies, with the vibrational effective temperature as a key controlling factor. From a simulation point of view, the absence of a vibrational thermostat (sometimes referred as the completely undamped limit in the context of inelastic nanoscale transport [29]) offers the advantage that a single real-time ECEID simulation, with $N$ allowed to respond and scan a range of values, probes a range of different temperature-dependent regimes.

In reality, there are two limiting mechanisms at play. On the one hand, the phonon mode will be coupled to the environment, typically through hydrogen-bonding to other water molecules. Therefore, part of this energy will be used in heating the aqueous environment. This energy transfer will correspond to a specific time scale related to the thermal conductivity of water, which is relatively low. If energy is pumped into this phonon mode at a faster rate, then it will accumulate in the bond. Here the harmonic approximation breaks down, and the inclusion of anharmonic potentials that incorporate the possibility of bond dissociation becomes essential. Such a resonant damage mechanism is reminiscent of dissociative electron attachment in DNA-related systems [30, 31], where electrons trapped in a resonant state transfer sufficient energy to a vibrational mode to break the corresponding bond. Incoming electrons with a well defined energy could trigger inelastic effects such as the ones described in the present chapter and be a prominent cause for dramatic heating and bond breaking.

## 8.2.5  Eigenstate Lifetime and Band Edge Trapping

In Fig. 8.17c,d we noticed that the $\delta$EEP decays slower for pulses above and below the FCB. This observation provides an indirect measure for the lifetime of the inelastically injected electrons in the water as a function of the initial injection energy. We know that electrons entering the water from these elastically forbidden energies do so by phonon emission and absorption, landing in the highest and lowest energy states in the FCB. If these states were long lived, they would act as an effective electron trap.

To understand this phenomenon, we evaluate the lifetimes of all eigenstates, $j$, in the water FCB, calculated in two different ways. The first is the purely elastic lifetime against tunnelling out into the leads given by $\tau_{\Sigma,j}^{-1} = -\frac{2}{\hbar}\text{Im}\langle j|\hat{\Sigma}_c^+(E_j)|j\rangle$ where $\hat{\Sigma}_c^+$ is the water chain's self-energy which incorporates the embedding in the environment through the end sites of the water chain that are coupled to the leads. The variations in $\tau_{\Sigma,j}^{-1}$, over the given narrow energy range, come mainly from the amplitudes of the water states $|j\rangle$ on the end sites of the chain. The second is directly from dynamical simulations, in which we release an excess electron from a chosen water eigenstate, propagate the system in time and fit an exponential decay to the decreasing EEP. We perform ECEID calculations for the purely elastic case and for the two earlier phonon occupancies $N(0) = 0$ and $N(0) = 2$. We include only the OB extraction term and, to concentrate on the lifetime, we keep $N$ frozen at its initial value. In the real-time simulations we further calculate and compare the lifetime associated with the population of the chosen initial state itself $\tau_{s,j}$ and the sum of populations of all FCB states $\tau_{a,j}$.

The characteristic horseshoe shape that all the decay times display in Fig. 8.19 fundamentally originates from the dominant elastic mechanism. It can be understood from the form of the self-energy and the chain eigenstates. More generally, of course, the lifetime against escaping into the environment will depend on the details of the system-environment coupling. However, as our specific example illustrates, one may in general expect different system excited states to have significantly different lifetimes against the environment. The self-energy lifetime $\tau_{\Sigma,j}$ tends to agree with $\tau_{s,j}$, especially in the low energy range, while $\tau_{a,j}$ is longer throughout the energy range. Any difference between $\tau_{s,j}$ and $\tau_{a,j}$ signals higher-order scattering processes, not captured by the perturbative $\tau_{\Sigma,j}$. Elastic ECEID calculations agree closely with the

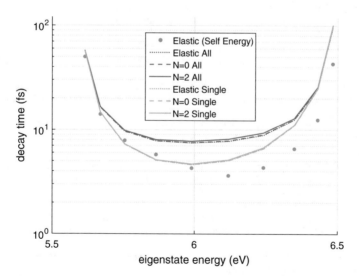

**Fig. 8.19** Lifetime comparison of the water FCB eigenstates for: static self energy calculations ($\tau_{\Sigma,j}$) (solid dots), exponential fits of the decaying population of a single initially populated level ($\tau_{s,j}$) (green curves), exponential fits of the total population of all excited states within the FCB, starting from a single initial populated FCB level ($\tau_{a,j}$) (red curves). Dotted curves correspond to elastic ECEID real-time calculations, dashed and solid lines to $N(0) = 0$ and $N(0) = 2$ ECEID simulations, both with $\dot{N}$ kept equal to 0 [9]

inelastic ones, indicating that the leading electron escape mechanism from the FCB is the elastic one to the leads, similarly to what was observed in earlier, perturbative inelastic tunnelling simulations [12]. The lifetimes computed here are comparable to the resonance lifetimes reported for tunnelling through 3d water structures in [11, 13].

The states at the band edge have a lifetime about one order of magnitude longer than the rest. They are the states most involved in the electron-phonon injection discussed earlier. A focused phonon-assisted electron injection in combination with the energy-dependent lifetimes of the FCB states, can thus provide an effective trapping mechanism for incident carriers, dependent on their incoming energy.

In Appendix C, we offer an alternative and simpler water chain model that reached similar results with some additional interesting features.

# References

1. Baccarelli, I., I. Bald, F.A. Gianturco, E. Illenberger, and J. Kopyra. 2011. Electron-induced damage of DNA and its components: Experiments and theoretical models. *Physics Reports* 508 (1–2): 1–44. https://doi.org/10.1016/j.physrep.2011.06.004.
2. Michael, B.D. 2000. A sting in the tail of electron tracks. *Science* 287 (5458): 1603–1604. https://doi.org/10.1126/science.287.5458.1603.
3. Alizadeh, E., T.M. Orlando, and L. Sanche. 2015. Biomolecular damage induced by ionizing radiation: The direct and indirect effects of low-energy electrons on DNA. *Annual Review of Physical Chemistry* 66: 379–98. https://doi.org/10.1146/annurev-physchem-040513-103605.
4. Pimblott, S.M., and J.A. LaVerne. 2007. Production of low-energy electrons by ionizing radiation. *Radiation Physics and Chemistry* 76 (8–9): 1244–1247. https://doi.org/10.1016/j.radphyschem.2007.02.012.
5. Boudaïffa, B., P. Cloutier, D. Hunting, M. Huels, and L. Sanche. 2000. Resonant formation of DNA strand breaks by low-energy (3 to 20 eV) electrons. *Science* 287 (5458): 1658–1660. https://doi.org/10.1126/science.287.5458.1658.
6. Martin, F., P. Burrow, Z. Cai, P. Cloutier, D. Hunting, and L. Sanche. 2004. DNA strand breaks induced by 0–4 eV electrons: The role of shape resonances. *Physical Review Letters* 93 (6): 068101. https://doi.org/10.1103/PhysRevLett.93.068101.
7. Cho, W., M. Michaud, and L. Sanche. 2004. Vibrational and electronic excitations of $H_2O$ on thymine films induced by low-energy electrons. *The Journal of Chemical Physics* 121 (22): 11289. https://doi.org/10.1063/1.1814057.
8. Simons, J. 2006. How do low-energy (0.1 − 2 eV) electrons cause DNA strand breaks? *Accounts of Chemical Research* 39 (10): 772–779. https://doi.org/10.1021/ar0680769.
9. Rizzi, V., T.N. Todorov, and J.J. Kohanoff. 2017. Inelastic electron injection in a water chain. *Scientific Reports* 7: 45410. https://doi.org/10.1038/srep45410.
10. Cooper, G.M. 2000. *The Cell: A Molecular Approach*, 2nd ed. Sunderland, MA: Sinauer Associates. ISBN 9780878931064.
11. Peskin, U., A. Edlund, I. Bar-On, M. Galperin, A. Nitzan, and Å. Edlund. 1999. Transient resonance structures in electron tunneling through water. *The Journal of Chemical Physics* 111 (16): 7558. https://doi.org/10.1063/1.480082.
12. Galperin, M., A. Nitzan, and U. Peskin. 2001. Traversal time for electron tunneling in water. *Journal of Chemical Physics* 114 (21): 9205–9208. https://doi.org/10.1063/1.1376162.
13. Galperin, M., and A. Nitzan. 2001. Inelastic effects in electron tunneling through water layers. *The Journal of Chemical Physics* 115 (6): 2681–2694. https://doi.org/10.1063/1.1383991.

14. Smyth M., and J. Kohanoff. 2011. Excess electron localization in solvated DNA bases. *Physical Review Letters* 106 (23): 238108. https://doi.org/10.1103/PhysRevLett.106.238108

15. McAllister, M., M. Smyth, B. Gu, G.A. Tribello, and J. Kohanoff. 2015. Understanding the interaction between low-energy electrons and DNA nucleotides in aqueous solution. *The Journal of Physical Chemistry Letters* 6 (15): 3091–3097. https://doi.org/10.1021/acs.jpclett.5b01011.

16. Paxton, A.T., and J.J. Kohanoff. 2011. A tight binding model for water. *The Journal of chemical physics* 134 (4): 044130. https://doi.org/10.1063/1.3523983.

17. Lide, D., and D. Tildesley. 2002. *CRC handbook of chemistry and physics*, 83rd ed. Boca Raton, Florida: Chemical Rubber Publishing Company.

18. Lide, D.R. (2005). *CRC handbook of chemistry and physics*, 85th ed. CRC press, no. v. 85. ISBN 9780849304859.

19. Todorov, T.N. 2002. Tight-binding simulation of current-carrying nanostructures. *Journal of Physics: Condensed Matter* 14 (11): 3049–3084. https://doi.org/10.1088/0953-8984/14/11/314.

20. Paulsson, M., T. Frederiksen, and M. Brandbyge. 2005. Modeling inelastic phonon scattering in atomic- and molecular-wire junctions. *Physical Review B–Condensed Matter and Materials Physics* 72 (20): 1–4. https://doi.org/10.1103/PhysRevB.72.201101.

21. Frederiksen, T., N. Lorente, M. Paulsson, and M. Brandbyge. 2007. From tunneling to contact: Inelastic signals in an atomic gold junction from first principles. *Physical Review B-Condensed Matter and Materials Physics* 75 (23): 1–8. https://doi.org/10.1103/PhysRevB.75.235441.

22. McEniry, E., T. Frederiksen, T. Todorov, D. Dundas, and A. Horsfield. 2008. Inelastic quantum transport in nanostructures: The self-consistent born approximation and correlated electron-ion dynamics. *Physical Review B* 78 (3): 035446. https://doi.org/10.1103/PhysRevB.78.035446.

23. X. Andrade, S. Hamel, and A. A. Correa. 2017. Non-linear conductivity of metals from real-time quantum simulations, pp. 1–8. arXiv: 1702.00411.

24. VandeVondele, J., M. Krack, F. Mohamed, M. Parrinello, T. Chassaing, and J. Hutter. 2005. Quickstep: Fast and accurate density functional calculations using a mixed Gaussian and plane waves approach. *Computer Physics Communications* 167 (2): 103–128. https://doi.org/10.1016/j.cpc.2004.12.014.

25. Goodwin, L., A.J. Skinner, and D.G. Pettifor. 1989. Generating transferable tight-binding parameters: Application to silicon. *Europhysics Letters (EPL)* 9 (7): 701–706. https://doi.org/10.1209/0295-5075/9/7/015.

26. Harrison, W. 1980. *Electronic structure and the properties of solids*. San Francisco: W.H. Freeman.

27. Fang, C., W.-F. Li, R.S. Koster, J. Klimeš, A. van Blaaderen, and M.A. van Huis. 2015. The accurate calculation of the band gap of liquid water by means of GW corrections applied to plane-wave density functional theory molecular dynamics simulations. *Physical Chemistry Chemical Physics* 17 (1): 365–375. https://doi.org/10.1039/C4CP04202F.

28. McEniry, E.J., T.N. Todorov, and D. Dundas. 2009. Current-assisted cooling in atomic wires. *Journal of Physics: Condensed Matter* 21 (19): 195304. https://doi.org/10.1088/0953-8984/21/19/195304.

29. Frederiksen, T., M. Paulsson, M. Brandbyge, and A.-P. Jauho. 2007. Inelastic transport theory from first principles: Methodology and application to nanoscale devices. *Physical Review B* 75 (20): 205413. https://doi.org/10.1103/PhysRevB.75.205413.

30. Haxton, D.J., Z. Zhang, H.-D. Meyer, T.N. Rescigno, and C.W. McCurdy. 2004. Dynamics of dissociative attachment of electrons to water through the 2B1 metastable state of the anion. *Physical Review A* 69 (6): 062714. https://doi.org/10.1103/PhysRevA.69.062714

31. Smyth, M., J. Kohanoff, and I.I. Fabrikant. 2014. Electron-induced hydrogen loss in uracil in a water cluster environment. *The Journal of Chemical Physics* 140 (18): 184313. https://doi.org/10.1063/1.4874841.

# Chapter 9
# A New Development: ECEID xp

This chapter presents a new methodological development that represents an improvement over ECEID. Through the use of a canonical transformation on the ECEID Hamiltonian, it is possible to include the dynamics of an extra classical degree of freedom in the model, a semiclassical oscillator position. Physically, we can think of it as a representation of the oscillator centroid, but it enters the derivation just as a time-dependent reference position. After deriving an exact set of equations, we compare it with an exact expansion from [1] and notice that, for a frozen centroid motion, exact ECEID from Sect. 5.2 is recovered. By applying approximations akin to ECEID, we reach a set of equations that contains the motion of the semiclassical oscillator position as a time-dependent parameter which still lacks an explicit equation of motion. We propose a condition that generates an Ehrenfest-like dynamics for the centroid motion and test it on a nanowire system. We then show a performance comparison between ECEID xp and ECEID. At last, we explore possible expansions of the method and further work. The work in this chapter was developed in collaboration with Alfredo Correa during a stay at Lawrence Livermore National Laboratory.

## 9.1 A Canonical Transformation

The derivation starts the same model Hamiltonian (5.1) used in ECEID

$$\hat{H} = \hat{H}_e + \sum_{\nu}^{N_o} \left( \frac{\hat{P}_{\nu}^2}{2M} + \frac{K\hat{X}_{\nu}^2}{2} - \hat{F}_{\nu}\hat{X}_{\nu} \right) \tag{9.1}$$

with $N_o$ harmonic oscillators coupled to electrons described by a general many-body electronic Hamiltonian $\hat{H}_e$. $\hat{X}_{\nu}$ and $\hat{P}_{\nu}$ are canonically conjugate operators, $[\hat{X}_{\nu}, \hat{P}_{\nu'}] = i\hbar\delta_{\nu\nu'}$ and the electron-phonon coupling $\hat{F}_{\nu}$ is an electron-only operator.

© Springer International Publishing AG, part of Springer Nature 2018
V. Rizzi, *Real-Time Quantum Dynamics of Electron-Phonon Systems*,
Springer Theses, https://doi.org/10.1007/978-3-319-96280-1_9

For simplicity, in this derivation, we pick identical spring constants $K$ and masses $M$ for all the oscillators. The final result can be trivially adjusted by employing oscillator-specific $K_\nu$ and $M_\nu$.

The final aim of the ECEID derivation was to develop a set of equations for the electronic DM $\hat{\rho}_e(t)$ and the mean oscillator occupations $N_\nu(t)$, connected by correlation operators $\hat{\mu}_\nu(t)$, $\hat{\lambda}_\nu(t)$ and the motion of auxiliary operators $\hat{C}_\nu^c$, $\hat{A}_\nu^c$, $\hat{C}_\nu^s$, $\hat{A}_\nu^s$. The dynamics of ECEID explicitly evolves $\hat{\rho}_e(t)$ and $N_\nu(t)$, but it does not take into account the semiclassical motion of the oscillator in its harmonic potential. There is no quantity that describes the time-dependent position of each oscillator in its potential.

The description of such a quantity is essential for linking ECEID to simulations where the real-space position of the atoms is relevant, such as Ehrenfest-TDDFT or MD. Connecting ECEID to those methods is no easy task and presents several difficulties, but would be a very rewarding achievement, as it would open a wide range of applications for ECEID. A crucial step forward is the introduction of an explicit oscillator position in ECEID. Does it make sense to speak of an oscillator classical position in ECEID's formalism or does ECEID imply that the oscillators are localized and do not move? To answer this question, we introduce additional time-dependent parameters by proposing the change of variables

$$\hat{X}_\nu = x_\nu(t) + \hat{\xi}_\nu \tag{9.2}$$

$$\hat{P}_\nu = M\dot{x}_\nu(t) + \hat{\pi}_\nu, \tag{9.3}$$

where we include the new canonically conjugate operators $[\hat{\xi}_\nu, \hat{\pi}_{\nu'}] = i\hbar\delta_{\nu\nu'}$ and classical oscillator positions $x_\nu(t)$. The idea behind this is that $\hat{\xi}_\nu$ and $\hat{\pi}_\nu$ represent quantum fluctuations around a hypothetical semiclassical trajectory described by $x_\nu(t)$, whose equation of motion is yet unspecified.

This transformation of phase-space variables is canonical and time-dependent, with a generating function of second-type [2]

$$G_2(X, \pi, t) = \sum_{\nu=1}^{N_o} \left( M\dot{x}_\nu(t)X_\nu + \pi_\nu X_\nu - x_\nu(t)\pi_\nu \right) \tag{9.4}$$

which links the Hamiltonian in the old set of variables $\hat{H}(\hat{X}, \hat{P})$ with the Hamiltonian in the transformed ones $\hat{h}(\hat{\xi}, \hat{\pi})$. The following equalities hold

$$\hat{P}_\nu = \frac{\partial G_2}{\partial X_\nu} \tag{9.5}$$

$$\hat{\xi}_\nu = \frac{\partial G_2}{\partial \pi_\nu} \tag{9.6}$$

$$\hat{h} = \hat{H} + \frac{\partial G_2}{\partial t}. \tag{9.7}$$

where the first two are simply the canonical transformation in Eqs. (9.2) and (9.3) and the last one defines an auxiliary equivalent Hamiltonian in the new variables

$$
\hat{h}(t) = \hat{H}_e + \sum_{\nu=1}^{N_o} \left( \frac{\hat{\pi}_\nu^2}{2M} + \frac{K\hat{\xi}_\nu^2}{2} + \frac{M\dot{x}_\nu(t)^2}{2} + \frac{Kx_\nu(t)^2}{2} + M\ddot{x}_\nu(t)x_\nu(t) \right.
$$
$$
\left. + (Kx_\nu(t) + M\ddot{x}_\nu(t))\hat{\xi}_\nu - \hat{F}_\nu \left( x_\nu(t) + \hat{\xi}_\nu \right) \right). \tag{9.8}
$$

Comparing Eq. (9.8) with (9.1), we see that operators $\hat{\xi}_\nu$ and $\hat{\pi}_\nu$ effectively replace $\hat{X}_\nu$ and $\hat{P}_\nu$ as the canonical displacement and momentum operators. Besides some scalar extra terms that depend on $x_\nu(t)$, $\dot{x}_\nu(t)$ and $\ddot{x}_\nu(t)$, a new term appears, $(Kx_\nu(t) + M\ddot{x}_\nu(t))\hat{\xi}_\nu$, which is linear in the quantum displacement. So far, the $x_\nu(t)$ trajectories are free parameters.

## 9.2 Exact Dynamics

In analogy with Chap. 5, the dynamics of the full DM follows the Liouville equation

$$
i\hbar\dot{\hat{\rho}}(t) = [\hat{h}(t), \hat{\rho}(t)] \tag{9.9}
$$

and the electronic DM is determined by

$$
i\hbar\dot{\hat{\rho}}_e(t) = [\hat{H}_e, \hat{\rho}_e(t)] - \sum_{\nu=1}^{N_o} \left( [\hat{F}_\nu, \hat{\mu}_\nu(t)] + x_\nu(t)[\hat{F}_\nu, \hat{\rho}_e(t)] \right). \tag{9.10}
$$

The derivation of the time-evolution for the mean oscillator occupation[1] $N_\nu(t) = \text{Tr}\left(\hat{a}_\nu^\dagger \hat{a}_\nu \hat{\rho}(t)\right)$ starts from the analogue of Eq. (5.9)

$$
\dot{N}_\nu(t) = \text{Tr}\left(\hat{N}_\nu \dot{\hat{\rho}}(t)\right) = \frac{1}{i\hbar}\text{Tr}\left([\hat{N}_\nu, \hat{h}(t)]\hat{\rho}(t)\right). \tag{9.11}
$$

The commutator above is

$$
\left[\hat{N}_\nu, \hat{h}(t)\right] = -\frac{i}{M\omega}\hat{\pi}_\nu\left(Kx_\nu(t) + M\ddot{x}_\nu(t) - \hat{F}_\nu\right), \tag{9.12}
$$

which makes the EOM for $\dot{N}_\nu(t)$

---

[1] Where now $\hat{a}_\nu^{(\dagger)}$ is the creation (annihilation) operator associated with the canonical displacement $\hat{\xi}_\nu$ and momentum $\hat{\pi}_\nu$ of oscillator $\nu$.

$$\dot{N}_\nu(t) = \frac{1}{\hbar M \omega} \left( \text{Tr}_\text{e} \left( \hat{F}_\nu \hat{\lambda}_\nu(t) \right) - \left( K x_\nu(t) + M \ddot{x}_\nu(t) \right) \bar{\pi}_\nu(t) \right). \tag{9.13}$$

In Eqs. (9.10) and (9.13) we introduced electronic correlation operators

$$\hat{\mu}_\nu(t) = \text{Tr}_0 \left( \hat{\xi}_\nu \, \hat{\rho}(t) \right) \tag{9.14}$$

$$\hat{\lambda}_\nu(t) = \text{Tr}_0 \left( \hat{\pi}_\nu \, \hat{\rho}(t) \right), \tag{9.15}$$

akin to the operators from (5.6) and (5.11). These operators store the correlated information between the electronic subsystem and oscillator and play an essential role in driving the EOM for $\hat{\rho}_\text{e}(t)$ and $N_\nu(t)$. The electronic traces of these operators

$$\bar{\xi}_\nu(t) = \text{Tr}_\text{e} \left( \hat{\mu}_\nu(t) \right) = \text{Tr} \left( \hat{\xi}_\nu \, \hat{\rho}(t) \right) \tag{9.16}$$

$$\bar{\pi}_\nu(t) = \text{Tr}_\text{e} \left( \hat{\lambda}_\nu(t) \right) = \text{Tr} \left( \hat{\pi}_\nu \, \hat{\rho}(t) \right) \tag{9.17}$$

correspond to the expectation value of $\hat{\xi}_\nu$ and $\hat{\pi}_\nu$.

In Chap. 5, the derivation proceeded to determine a time-local form of $\hat{\mu}_\nu(t)$ and $\hat{\lambda}_\nu(t)$ by inserting in their definition an exact integral form of $\hat{\rho}_\text{e}(t)$ (5.4). Here we follow a different path and explicitly evaluate the time derivative of the correlation operators.[2] Using Eq. (9.9) in the time derivative of the correlation operators, we have

$$i\hbar \dot{\hat{\mu}}_\nu(t) = \text{Tr}_0 \left( \hat{\xi}_\nu [\hat{h}, \hat{\rho}(t)] \right) \tag{9.18}$$

$$i\hbar \dot{\hat{\lambda}}_\nu(t) = \text{Tr}_0 \left( \hat{\pi}_\nu [\hat{h}, \hat{\rho}(t)] \right). \tag{9.19}$$

Employing permutations under the trace and the canonical commutation relations, the equations above can be written as[3]

---

[2]This alternative approach is also applicable to ECEID in Chap. 5. There, the formalism with the auxiliary operators was chosen for consistency with the method as it was originally developed and presented in [3, 4]. Here, the derivation follows a sleeker approach that does not involve the definition of any auxiliary operators. The dynamics resulting from the two approaches is equivalent.

[3]For example, the contribution of the electron-phonon coupling term $-\hat{F}_\nu \hat{\xi}_\nu$ in Eq. (9.19) is

$$-\sum_{\nu'=1}^{N_o} \text{Tr}_0 \left( \hat{\pi}_\nu \hat{F}_{\nu'} \hat{\xi}_{\nu'} \hat{\rho}(t) - \hat{\pi}_\nu \hat{\rho}(t) \hat{F}_{\nu'} \hat{\xi}_{\nu'} \right) = -\sum_{\nu'=1}^{N_o} \text{Tr}_0 \left( \hat{\pi}_\nu \hat{\xi}_{\nu'} \hat{F}_{\nu'} \hat{\rho}(t) - \hat{\xi}_{\nu'} \hat{\pi}_\nu \hat{\rho}(t) \hat{F}_{\nu'} \right) =$$

$$-\sum_{\nu'=1}^{N_o} \left( \text{Tr}_0 \left( \hat{\pi}_\nu \hat{\xi}_{\nu'} [\hat{F}_{\nu'}, \hat{\rho}(t)] \right) - i\hbar \delta_{\nu\nu'} \text{Tr}_0 \left( \hat{\rho}(t) \hat{F}_{\nu'} \right) \right).$$

$$\tag{9.20}$$

$$i\hbar\dot{\hat{\mu}}_\nu(t) = \left[\hat{H}_e, \hat{\mu}_\nu(t)\right] + \frac{i\hbar}{M}\hat{\lambda}_\nu(t) - \sum_{\nu'=1}^{N_o}\left(\left[\hat{F}_{\nu'}, \text{Tr}_o(\hat{\xi}_\nu\hat{\xi}_{\nu'}\hat{\rho}(t))\right] + \left(x_{\nu'}(t)[\hat{F}_{\nu'}, \hat{\mu}_\nu(t)]\right)\right)$$

$$(9.21)$$

$$i\hbar\dot{\hat{\lambda}}_\nu(t) = \left[\hat{H}_e, \hat{\lambda}_\nu(t)\right] - i\hbar K\hat{\mu}_\nu(t) - \sum_{\nu'=1}^{N_o}\left(\left[\hat{F}_{\nu'}, \text{Tr}_o(\hat{\pi}_\nu\hat{\xi}_{\nu'}\hat{\rho}(t))\right] - i\hbar\hat{\rho}_e(t)\hat{F}_\nu\right.$$

$$\left. + x_{\nu'}(t)[\hat{F}_{\nu'}, \hat{\lambda}_\nu(t)]\right) - i\hbar\left(M\ddot{x}_\nu(t) + Kx_\nu(t)\right)\hat{\rho}_e(t)$$

$$(9.22)$$

which are exact.

Equations (9.21) and (9.22) contain second order correlation terms $\text{Tr}_o(\hat{\xi}_\nu\hat{\xi}_{\nu'}\hat{\rho}(t))$ and $\text{Tr}_o(\hat{\pi}_\nu\hat{\xi}_{\nu'}\hat{\rho}(t))$. These terms can be either approximated, as we did in ECEID and we do here in Sect. 9.3, or their time evolution has to be evaluated explicitly. They would have to be defined as second order correlation operators and require approximations on the third order correlations. The procedure would escalate the EOM to the next order, carrying truncation difficulties and significantly reducing the computational efficiency.

The second order operators depend on two indices $\nu\nu'$ and their number scales as $N_o^2$. The computational cost of a method that includes them would clearly not scale linearly with the number of oscillators. Moreover, the ECEID approximations on the second order terms determine a truncation that is physically meaningful. A physically consistent truncation strategy on the third order terms is trickier to figure out.

This expansion strategy is very reminiscent of the one used in CEID [1, 5]. There, the atomic potential is general and the method is derived by expanding quantities around an average atomic position for small displacements. By plugging an harmonic potential into CEID and evaluating the exact expansion from Eq. (8) in [1], it is possible to derive exact equations for the correlation operators that here we call $\hat{\mu}_\nu(t)$ and $\hat{\lambda}_\nu(t)$. It can be verified that the resulting expressions correspond term by term with the ECEID xp exact results from Eqs. (9.21) and (9.22). The fact that the two strategies reach the same conclusion is reassuring and supports the canonical transformation strategy employed here.

## 9.3 The Approximations and ECEID xp

The second order terms in Eqs. (9.21) and (9.22) can be written as

$$\text{Tr}_o\left(\hat{\xi}_\nu\hat{\xi}_{\nu'}\hat{\rho}(t)\right) = \frac{\hbar}{2M\omega}\text{Tr}_o\left((\hat{a}_\nu\hat{a}_{\nu'} + \hat{a}_\nu\hat{a}_{\nu'}^\dagger + \hat{a}_\nu^\dagger\hat{a}_{\nu'} + \hat{a}_\nu^\dagger\hat{a}_{\nu'}^\dagger)\hat{\rho}(t)\right) \quad (9.23)$$

$$\text{Tr}_o\left(\hat{\pi}_\nu\hat{\xi}_{\nu'}\hat{\rho}(t)\right) = \frac{i\hbar}{2}\text{Tr}_o\left((\hat{a}_\nu^\dagger\hat{a}_{\nu'} - \hat{a}_\nu\hat{a}_{\nu'} - \hat{a}_\nu\hat{a}_{\nu'}^\dagger + \hat{a}_\nu^\dagger\hat{a}_{\nu'}^\dagger)\hat{\rho}(t)\right) \quad (9.24)$$

by using the second quantization form of $\hat{\xi}_\nu$ and $\hat{\pi}_\nu$.

Just as in ECEID, the approximations consists of ignoring all cross terms that involve different oscillators $\text{Tr}_o\left(\hat{a}_\nu^{(\dagger)}\hat{a}_{\nu'}^{(\dagger)}\hat{\rho}(t)\right) \simeq \text{Tr}_o\left(\hat{a}_\nu^{(\dagger)}\hat{a}_{\nu'}^{(\dagger)}\hat{\rho}(t)\right)\delta_{\nu\nu'}$ and terms with double (de)excitations[4] $\text{Tr}_o\left(\hat{a}_\nu\hat{a}_\nu\hat{\rho}(t)\right) = \text{Tr}_o\left(\hat{a}_\nu^\dagger\hat{a}_\nu^\dagger\hat{\rho}(t)\right) \simeq 0$. With these approximations and $[\hat{a}_\nu, \hat{a}_\nu^\dagger] = 1$, Eqs. (9.23)–(9.24) become

$$\text{Tr}_o\left(\hat{\xi}_\nu\hat{\xi}_\nu\hat{\rho}(t)\right) \simeq \frac{\hbar}{M\omega}\text{Tr}_o\left(\left(\hat{a}_\nu^\dagger\hat{a}_\nu + \frac{1}{2}\right)\hat{\rho}(t)\right) \tag{9.25}$$

$$\text{Tr}_o\left(\hat{\pi}_\nu\hat{\xi}_\nu\hat{\rho}(t)\right) \simeq -\frac{i\hbar}{2}\hat{\rho}_e(t) \tag{9.26}$$

Equation (9.26) is ready to be inserted back in Eq. (9.22). Its form is reminiscent of an uncertainty principle on the the displacement and momentum operators. Equation (9.25) cannot yet be used because $\text{Tr}_o\left(\hat{a}_\nu^\dagger\hat{a}_\nu\hat{\rho}(t)\right)$ is an unknown electronic operator that, in principle, needs its own EOM. In (9.25) (but not earlier) we split the DM into $\hat{\rho}(t) \simeq \hat{\rho}_e(t)\hat{\rho}_o(t)$ so that Eq. (9.25) can be written as

$$\text{Tr}_o\left(\hat{\xi}_\nu\hat{\xi}_\nu\hat{\rho}(t)\right) \simeq \frac{\hbar}{M\omega}\left(N_\nu(t) + \frac{1}{2}\right)\hat{\rho}_e(t), \tag{9.27}$$

that now can be included in Eq. (9.21).

The final approximated EOM for the correlation operators are

$$\dot{\hat{\mu}}_\nu(t) = \frac{1}{i\hbar}\left[\hat{H}_e, \hat{\mu}_\nu(t)\right] + \frac{1}{M}\hat{\lambda}_\nu(t) + \frac{i}{M\omega}\left(N_\nu(t) + \frac{1}{2}\right)\left[\hat{F}_\nu, \hat{\rho}_e(t)\right] - \frac{1}{i\hbar}\sum_{\nu'=1}^{N_o}\left(x_{\nu'}(t)\left[\hat{F}_{\nu'}, \hat{\mu}_\nu(t)\right]\right) \tag{9.28}$$

$$\dot{\hat{\lambda}}_\nu(t) = \frac{1}{i\hbar}\left[\hat{H}_e, \hat{\lambda}_\nu(t)\right] - K\hat{\mu}_\nu(t) + \frac{1}{2}\left\{\hat{F}_\nu, \hat{\rho}_e(t)\right\} - \left(M\ddot{x}_\nu(t) + Kx_\nu(t)\right)\hat{\rho}_e(t)$$
$$- \frac{1}{i\hbar}\sum_{\nu'=1}^{N_o}\left(x_{\nu'}(t)\left[\hat{F}_{\nu'}, \hat{\lambda}_\nu(t)\right]\right), \tag{9.29}$$

where the new terms are indicated in blue. Equations (9.28) and (9.29), together with (9.10) and (9.13), represent the EOM of the methodological development called ECEID xp.[5]

---

[4]In Appendix D, a recent development is shown where the double (de)excitations approximation is not invoked.

[5]$\hat{\mu}_\nu(t)$ and $\hat{\lambda}_\nu(t)$ in Eqs. (9.28) and (9.29) contain terms summed over $\nu'$. Therefore, in ECEID xp, the correlation operators depend on all the oscillators. The parallelism of the method hinges on the fact that every operator associated with oscillator $\nu$ depends only on purely electronic quantities and on oscillator-$\nu$-specific quantities. The implementation of the new formalism would reduce the effectiveness of the parallelism in the presence of many oscillators. A possible solution, consistent

It is important to notice that, in the limit of $x_\nu(t) = 0$ for all times, ECEID's EOM from Chap. 5 are recovered. To prove this, it suffices to take the time derivative of Eqs. (5.34) and (5.35) and use Eqs. (5.36) and (5.39). This limit is a crucial feature of the new formulation because it allows to connect ECEID to ECEID xp by switching a parameter on and off.

One key element is missing before the ECEID xp set of equations can be used in simulations: an EOM for $x_\nu(t)$. Up to this point, ECEID xp is general and $x_\nu(t)$ is an as yet unspecified time-dependent parameter. Ideally its EOM would be derived from a minimization condition such as the principle of least action. We explore a possible condition in Sect. 9.4 and test it on a nanowire.

## 9.4 An Ehrenfest-Like Condition for $x_\nu(t)$

The introduction of $x_\nu(t)$ in the method had the aim to describe the oscillator position semiclassically and no initial assumption was made on its dynamics. Now, in ECEID xp, $x_\nu(t)$ appears in the EOM, but it still needs an EOM itself so that the full set of equations can be finally simulated. A possibility to determine its dynamics is to impose a condition on the ECEID xp approximated terms. For example, in Eq. (9.26), one can impose that the total trace is

$$\text{Tr}\left(\hat{\xi}_\nu \hat{\pi}_\nu \hat{\rho}(t)\right) \simeq \frac{i\hbar}{2} \tag{9.30}$$

at all times, or equivalently

$$\frac{\text{dTr}\left(\hat{\xi}_\nu \hat{\pi}_\nu \hat{\rho}(t)\right)}{\text{d}t} \simeq 0. \tag{9.31}$$

This condition can be interpreted as keeping the ECEID approximation valid all along the time evolution of $x_\nu(t)$.

By using the Liouville equation (9.9) and the canonical commutation relations, the condition (9.31) can be written as

$$\frac{1}{i\hbar}\text{Tr}\left([\hat{\xi}_\nu \hat{\pi}_\nu, \hat{h}(t)]\hat{\rho}(t)\right) = -K\text{Tr}\left(\hat{\xi}_\nu^2 \hat{\rho}(t)\right) + \frac{1}{M}\text{Tr}\left(\hat{\pi}_\nu^2 \hat{\rho}(t)\right) \tag{9.32}$$

$$-\left(M\ddot{x}_\nu(t) + Kx_\nu(t)\right)\text{Tr}\left(\hat{\xi}_\nu \hat{\rho}(t)\right) + \text{Tr}\left(\hat{F}_\nu \hat{\xi}_\nu \hat{\rho}(t)\right) \simeq 0. \tag{9.33}$$

With the use of Eq. (9.25), of

$$\text{Tr}\left(\hat{\pi}_\nu^2 \hat{\rho}(t)\right) \simeq \hbar M\omega \left(N_\nu(t) + \frac{1}{2}\right) \tag{9.34}$$

---

with the spirit of ECEID's approximations is to insert a $\delta_{\nu\nu'}$ in Eqs. (9.28) and (9.29) and consider only operators depending on $\nu$.

and the DM decoupling $\hat{\rho}(t) \simeq \hat{\rho}_e(t)\hat{\rho}_o(t)$, the condition (9.31) gives

$$\left(M\ddot{x}_\nu(t) + Kx_\nu(t) - \mathrm{Tr}_e(\hat{F}_\nu\hat{\rho}_e(t))\right)\bar{\xi}_\nu(t) = 0. \tag{9.35}$$

In general, $\bar{\xi}_\nu(t)$ is not zero, therefore it is necessary that

$$M\ddot{x}_\nu(t) = -Kx_\nu(t) + \mathrm{Tr}_e\left(\hat{F}_\nu\hat{\rho}_e(t)\right) \tag{9.36}$$

is valid.

Equation (9.36) is an EOM for $x_\nu(t)$ and is reminiscent of Ehrenfest dynamics, however its derivation did not invoke the Ehrefenst theorem. The similarity with Ehrenfest dynamics allows us to identify $x_\nu(t)$ with a semiclassical trajectory, mainly in the regimes where the $N_\nu(t)$ is small. For large $N_\nu(t)$, the position of the oscillator would be dominated by the quantum uncertainty and $x_\nu(t)$ would represent an average trajectory over quantum and ensemble states.

By including the Ehrenfest-like condition from Eq. (9.36) in the ECEID xp EOM, the following closed set of equations is determined

$$i\hbar\dot{\hat{\rho}}_e(t) = [\hat{H}_e, \hat{\rho}_e(t)] - \sum_{\nu=1}^{N_o}\left([\hat{F}_\nu, \hat{\mu}_\nu(t)] + x_\nu(t)[\hat{F}_\nu, \hat{\rho}_e(t)]\right) \tag{9.37a}$$

$$\hbar M\omega\dot{N}_\nu(t) = \mathrm{Tr}_e\left(\hat{F}_\nu\hat{\lambda}_\nu(t)\right) - \mathrm{Tr}_e\left(\hat{F}_\nu\hat{\rho}_e(t)\right)\bar{\pi}_\nu(t) \tag{9.37b}$$

$$i\hbar\dot{\hat{\mu}}_\nu(t) = [\hat{H}_e, \hat{\mu}_\nu(t)] + \frac{i\hbar}{M}\hat{\lambda}_\nu(t) - \frac{\hbar}{M\omega}\left(N_\nu(t) + \frac{1}{2}\right)[\hat{F}_\nu, \hat{\rho}_e(t)]$$
$$- \sum_{\nu'=1}^{N_o}\left(x_{\nu'}(t)[\hat{F}_{\nu'}, \hat{\mu}_\nu(t)]\right) \tag{9.37c}$$

$$i\hbar\dot{\hat{\lambda}}_\nu(t) = [\hat{H}_e, \hat{\lambda}_\nu(t)] - i\hbar K\hat{\mu}_\nu(t) + \frac{i\hbar}{2}\{\hat{F}_\nu, \hat{\rho}_e(t)\} - i\hbar\mathrm{Tr}_e\left(\hat{F}_\nu\hat{\rho}_e(t)\right)\hat{\rho}_e(t)$$
$$- \sum_{\nu'=1}^{N_o}\left(x_{\nu'}(t)[\hat{F}_{\nu'}, \hat{\lambda}_\nu(t)]\right) \tag{9.37d}$$

$$M\ddot{x}_\nu(t) = -Kx_\nu(t) + \mathrm{Tr}_e\left(\hat{F}_\nu\hat{\rho}_e(t)\right) \tag{9.37e}$$

where all the electronic operators are many-body operators and the new terms that characterize ECEID xp are indicated in blue.

The Eq. (9.37a) can be integrated in time, after initial conditions are chosen by usual methods. In practice, another approximation is needed to make the problem numerically tractable: making the electronic problem one-body. The one-body projection procedure is analogous to the one for ECEID in Sect. 5.5 and it acts on the anticommutator term in Eq. (9.37d).

The many-body anticommutator transforms into

$$\left\{ \hat{F}_\nu, \hat{\rho}_e(t) \right\} - 2\hat{\rho}_e(t) \hat{F}_\nu \hat{\rho}_e(t) + 2\hat{\rho}_e(t) \, \mathrm{Tr}_e\left( \hat{F}_\nu \hat{\rho}_e(t) \right) \tag{9.38}$$

with one-body operators, including the extra term that was ignored in Eq. (5.44) because it was related to the centroid motion. Here the centroid motion is essential and that term cannot be ignored. In fact, using Eq. (5.44) and writing the one-body form of $\dot{\hat{\lambda}}_\nu(t)$, we see that the extra term exactly cancels with another term in (9.37d), giving

$$i\hbar\dot{\hat{\lambda}}_\nu(t) = [\hat{H}_e, \hat{\lambda}_\nu(t)] - i\hbar K \hat{\mu}_\nu(t) + \frac{i\hbar}{2}\left\{ \hat{F}_\nu, \hat{\rho}_e(t) \right\} - i\hbar\hat{\rho}_e(t)\hat{F}_\nu\hat{\rho}_e(t)$$
$$- \sum_{\nu'=1}^{N_o}\left( x_{\nu'}(t)\left[ \hat{F}_{\nu'}, \hat{\lambda}_\nu(t) \right] \right) \tag{9.39}$$

where all terms are one-body. Moreover, the definitions of $\bar{\xi}_\nu(t)$ and $\bar{\pi}_\nu(t)$ in Eqs. (9.16) and (9.17) in the one-body picture have to be adjusted by including a prefactor of $1/N_e$.

By replacing the variables (9.2) in the original Hamiltonian (9.1) and taking the total trace, we obtain the following quantity, that we identify with the total energy of the system

$$E = \mathrm{Tr}_e\left( \hat{H}_e\hat{\rho}_e \right) + \sum_{\nu=1}^{N_o}\left( \hbar\omega\left( N_\nu(t) + \frac{1}{2} \right) + \frac{M\dot{x}_\nu(t)^2}{2} + \frac{Kx_\nu(t)^2}{2} \right.$$
$$\left. + Kx_\nu(t)\bar{\xi}_\nu(t) + M\dot{x}_\nu(t)\bar{\pi}_\nu(t) - x_\nu(t)\mathrm{Tr}_e\left( \hat{F}_\nu\hat{\rho}_e(t) \right) - \mathrm{Tr}_e\left( \hat{F}_\nu\hat{\mu}_\nu(t) \right) \right) \tag{9.40}$$

Tests confirmed that the total energy is conserved.

## 9.5 A Test Case

There are a few trivial limiting cases that can be verified by inspection. For instance, if $\hat{F}_\nu = 0$, the EOM of $x_\nu(t)$ and $\hat{\rho}_e(t)$ are decoupled and $N_\nu(t)$ becomes a constant of the motion. For small $x_\nu(t)$, the equations would effectively reduce to the ECEID ones. By setting the correlation operators $\hat{\mu}_\nu$ and $\hat{\lambda}_\nu$ to zero at all times, the EOM of $\hat{\rho}_e$ and $x_\nu$ reduce to those of Ehrenfest dynamics, eliminating important physics. In the following test, we call this latter limit classical Ehrenfest.

Initial conditions $N_\nu(0) = 0$ and $x_\nu(0) \neq 0$ (or $\dot{x}_\nu(0) \neq 0$) represent non-trivial semiclassical states of the oscillator. These states have the minimum achievable quantum uncertainty in position and momentum and they are the closest to a classical initial condition. During the ensuing dynamics, the classical description stays valid as long as the position uncertainty determined by $N_\nu(t)$ is small compared with the amplitude of motion of $x_\nu(t)$. To measure the quantum spread associated with $N_\nu(t)$, we can define

$$\Delta\xi_\nu(t) \equiv \sqrt{\left\langle \hat{\xi}_\nu^2(t) \right\rangle} = \sqrt{\left( N_\nu(t) + \frac{1}{2} \right) \frac{\hbar}{M\omega}} \tag{9.41}$$

which is a characteristic quantum spatial amplitude around $x_\nu(t)$ that is associated with the ensemble uncertainty. As $N_\nu(t)$ grows, the classical position and momentum of the oscillator become more and more uncertain. Since $x_\nu(t)$ follows Ehrenfest-like equations here, we expect that in closed systems it would eventually decay to small values (see Sect. 4.2).

The oscillator centroid is defined as the expectation value of the position operator $\hat{X}_\nu$ and after the canonical transformation can be written as

$$\bar{x}_\nu(t) = \text{Tr}\left( \hat{X}_\nu \hat{\rho}(t) \right) = x_\nu(t) + \bar{\xi}_\nu(t). \tag{9.42}$$

For small $\bar{\xi}_\nu(t)$, we can identify the semiclassical position $x_\nu(t)$ with the oscillator centroid $\bar{x}_\nu(t)$. The oscillator position in time can then be tracked by its centroid motion, with the uncertainty given by the quantum spread $\bar{x}_\nu(t) \pm \Delta\xi_\nu(t)$.

As a test of the method, we choose a system similar to the ones used in the Joule heating Sect. 6.4 and simulate it starting from the above semiclassical initial condition. A single oscillator is embedded in the middle of a perfect wire with zero onsite energies and 1 eV hoppings. The central region of the wire is made of 11 sites. The leads have 32 sites and present a $\Gamma = 0.8$ eV which appears in the Hamiltonian of Eqs. (9.37c) and (9.37d), as described in Sect. 6.2 for CEID's auxiliary operators. $\Gamma$ is applied to the 12 leftmost and rightmost sites in the left and right lead respectively. The oscillator has mass $M = 0.5$ a.m.u., $\hbar\omega = 0.2$ eV, $F = 1$ eV/Å, $N(0) = 0$ and $x_\nu(0) = 0.4$ Å. The initial electronic temperature is 10000 K.

In Fig. 9.1, we follow both the time-evolution of $x_1(t)$ and the quantum spread $\Delta\xi_1(t)$ around it. The dynamics of $x_1(t)$ displays quasiclassical oscillations at the beginning. During the dynamics, $\bar{\xi}_1$ and $\bar{\pi}_1$ remain negligibly small, confirming the assumption that $x_1(t)$ mimics the motion of the centroid. As the oscillator interacts with the electrons, the oscillations in $x_1(t)$ are damped. However, simultaneously with the damping, the spread increases, approaching a point of equilibrium for long times, as shown in Fig. 9.2a. At long times, the variable $x_1(t)$ is almost completely damped and effectively loses its meaning as a representative of the centre of mass motion, as the dominant oscillation mechanism lies in the quantum spread.

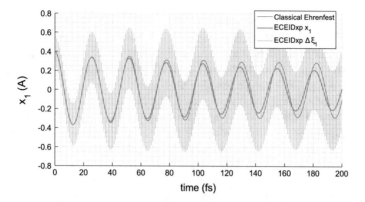

**Fig. 9.1** In blue, it is shown the dynamics of $x_1(t)$ during an ECEID xp simulation of a single oscillator in a perfect nanowire with $N_L = 32$ starting from $N(0) = 0$ and $x_1(0) = 0.4$ Å. The quantum spread $\xi_1(t)$ is displayed as grey vertical bars at fixed time intervals around the semiclassical position $x_1(t)$. The trajectory of $x_1(t)$ for an analogous classical Ehrenfest simulation is shown in red

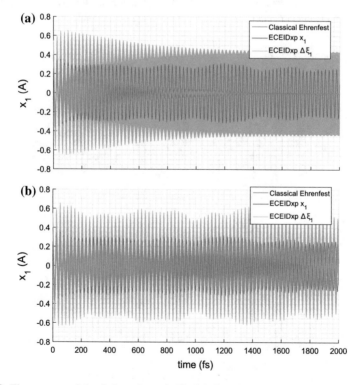

**Fig. 9.2** The same set of simulations shown in Fig. 9.1, for a longer time of 2 ps. In **a** $\Gamma = 0.8$ eV, in **b** $\Gamma = 0$

The ECEID xp results are compared to a classical Ehrenfest simulation, where both $\hat{\mu}_\nu$ and $\hat{\lambda}_\nu$ are turned off. The semiclassical oscillator motion displays damped oscillations just as in ECEID xp, with a slightly lower frequency and a less pronounced damping. At long times, the classical Ehrenfest simulation stops damping and settles, oscillating between about $-0.3\,\text{Å} < x_1(t) < 0.3\,\text{Å}$.

To understand the cause of the finite oscillations survival, we perform the same ECEID xp simulation with $\Gamma = 0$ and show the results in Fig. 9.2b. Without a $\Gamma$ to smear out the energy levels and favour the electron-oscillator energy transfer, ECEID xp displays a very similar dynamics when compared to classical Ehrenfest. In small systems, the absence of $\Gamma$ inhibits inelastic transitions in ECEID xp.

For larger systems, the requirement of a level broadening mechanism to trigger inelastic transitions becomes less rigid as the systems energy level spacing gets smaller. We test the same configuration as above with much longer leads $N_L = 200$ sites and show the results in Fig. 9.3. In (a), ECEID xp features $\Gamma = 0.12$ eV applied to the 180 leftmost and rightmost sites in the leads, while in (b) there is no $\Gamma$. The centroid dynamics $x_1(t)$ here is comparable between ECEID xp and classical Ehrenfest. In both cases, with and without $\Gamma$, it tends to dampen for long times,

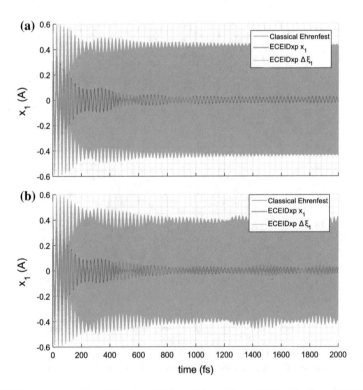

**Fig. 9.3** The same set of simulations as in Fig. 9.2 for a longer chain with $N_L = 200$ sites. In **a** $\Gamma = 0.12$ eV, in **b** $\Gamma = 0$

indicating an energy transfer away from the semiclassical degree of freedom, with a key difference that distinguishes ECEID xp from Ehrenfest. In the classical Ehrenfest simulation, the transfer of energy goes directly into the electrons, while in ECEID xp energy can go to the quantum spread $\xi_1(t)$.

The presence of electron-phonon correlation through $\hat{\mu}_\nu$ and $\hat{\lambda}_\nu$ allows the excess energy in $x_1(t)$ to be deposited into the oscillator spread. This mechanism is absent in classical Ehrenfest, where the energy can only go to the electrons. The possibility to capture the correlated energy exchange between $x_1(t)$ and $\xi_1(t)$ is a fundamental feature of ECEID xp.

## 9.6 Code Performance

The ECEID xp EOM with $x_\nu(t) = 0$ are mathematically equivalent to ECEID and they are more lightweight to simulate, as they present the explicit time evolution of only 2 operators for every oscillator against the 4 auxiliary operators in ECEID. To measure the performance improvements, the new formalism is implemented in the *ElPh* code and the same set of tests as the one in Sect. 6.5 is executed.

Table 9.1 shows the computational times, the oscillator and electron scaling and the percentage improvement against the previous implementation. The scalings are largely unaffected by the new development: they follow a similar pattern as the ECEID one in Sect. 6.5. The similarity is due to the fact that the basic algorithm and its optimizations are essentially unchanged.

The main change is the number of operators that explicitly evolve and, therefore, the number of matrix operations that are computed in every timestep. The new set of EOM is smaller in size and therefore faster to simulate. The computational times display remarkable improvements compared to before: the average decrease in computational time is 40%, with best-case scenario improvements of over 100% and worst-case scenario of about 20%. The better performance of the new implementation of the code is a welcome additional feature that makes the ECEID xp formulation with $x_\nu(t) = 0$ the method of choice for ECEID simulation and a valid stepping stone for future developments.

## 9.7 Final Remarks

The ECEID xp method presented in this chapter is a promising expansion of ECEID. Its limit $x_\nu(t) = 0$ where an equivalent and faster version of ECEID is recovered represents both a validation of the approach and a good starting point on which to build improvements. A recent development where the double (de)excitation approximation

**Table 9.1**  Performance test of the *ElPh* code with the ECEID xp set of EOM and $x_\nu(t) = 0$. Just as in Table 6.2, the computational time for simulations of $10^6$ timesteps is compared among systems with a varying number of oscillators and electronic sites. Oscillator and electronic scalings are shown following the colour convention introduced in Table 6.2 with green for good results, yellow for acceptable, red for poor. A percentage improvement over the previous implementation of the code is presented to measure the improvement

| Number of e. sites | Number of oscillators | Computational time | Oscillator scaling | Electronic scaling | Improvement versus ECEID (%) |
|---|---|---|---|---|---|
| 50 | 1 | 2 m 13 s | | | |
| | 2 | 2 m 32 s | 1.1× | | 44 |
| | 5 | 3 m 7 s | 1.4× | | 36 |
| | 10 | 4 m 6 s | 1.8× | | 27 |
| | 15 | 4 m 54 s | 2.2× | | 18 |
| | 20 | 6 m 0 s | 2.7× | | 20 |
| | 40 | 14 m 53 s | 6.7× | | 127 |
| 100 | 1 | 9 m 16 s | | 4.2× | 42 |
| | 2 | 10 m 11 s | 1.1× | 4.0× | 37 |
| | 5 | 12 m 20 s | 1.3× | 4.0× | 31 |
| | 10 | 16 m 10 s | 1.7× | 3.9× | 28 |
| | 15 | 20 m 10 s | 2.2× | 4.1× | 41 |
| | 20 | 25 m 1 s | 2.7× | 4.2× | 83 |
| | 40 | 1 h 21 m 35 s | 8.8× | 5.5× | 52 |
| 150 | 1 | 21 m 57 s | | 9.9× | 42 |
| | 2 | 23 m 37 s | 1.1× | 9.3× | 38 |
| | 5 | 29 m 0 s | 1.3× | 9.3× | 40 |
| | 10 | 43 m 28 s | 2.0× | 10.6× | 68 |
| | 15 | 1 h 10 m 30 s | 3.2× | 14.4× | 66 |
| | 20 | 1 h 44 m 0 s | 4.7× | 17.3× | 40 |
| | 40 | 3 h 41 m 48 s | 10.1× | 14.9× | 28 |
| 200 | 1 | 42 m 49 s | | 19.3× | 41 |
| | 2 | 46 m 38 s | 1.1× | 18.4× | 40 |
| | 5 | 1 h 0 m 50 s | 1.4× | 19.5× | 61 |
| | 10 | 1 h 42 m 0 s | 2.4× | 24.9× | 55 |
| | 15 | 2 h 42 m 10 s | 3.8× | 33.1× | 36 |
| | 20 | 4 h 9 m 20 s | 5.8× | 41.6× | 25 |
| | 40 | 7 h 15 m 40 s | 10.2× | 29.3× | 32 |
| 500 | 1 | 7 h 10 m 8 s | | 194.0 | 39 |
| | 2 | 7 h 23 m 27 s | 1.0 | 175.0 | 44 |
| | 5 | 10 h 53 m 0 s | 1.5 | 209.5 | 42 |
| | 10 | 16 h 35 m 20 s | 2.3× | 242.8× | 22 |
| | 15 | 1 d 0 h 19 m 30 s | 3.4× | 297.9× | 50 |
| | 20 | 1 d 10 h 46 m 0 s | 4.8× | 347.7× | 40 |
| | 40 | 2 d 12 h 37 m 0 s | 8.5× | 244.4× | 46 |

is not applied is presented in Appendix D. The Ehrenfest-like condition proposed and tested here is a possibility, but it is not necessarily the best choice. The search for a suitable dynamics of $x_\nu(t)$ is ongoing and is a fundamental future progress.

# References

1. Horsfield, A.P., D.R. Bowler, A.J. Fisher, T.N. Todorov, and C.G. Sanchez. 2005. Correlated electron-ion dynamics: The excitation of atomic motion by energetic electrons. *Journal of Physics: Condensed Matter* 17 (30): 4793–4812. https://doi.org/10.1088/0953-8984/17/30/006.
2. Goldstein, H., C. Poole, and J. Safko. 2002. *Classical mechanics*. Addison Wesley. ISBN 9780201657029.
3. Rizzi, V., T.N. Todorov, J.J. Kohanoff, and A.A. Correa. 2016. Electron-phonon thermalization in a scalable method for real-time quantum dynamics. *Physical Review B* 93 (2): 024306. https://doi.org/10.1103/PhysRevB.93.024306.
4. Rizzi, V., T.N. Todorov, and J.J. Kohanoff. 2017. Inelastic electron injection in a water chain. *Scientific Reports* 7: 45410. https://doi.org/10.1038/srep45410.
5. Horsfield, A.P., D.R. Bowler, A.J. Fisher, T.N. Todorov, and C.G. Sánchez. 2004. Beyond Ehrenfest: Correlated non-adiabatic molecular dynamics. *Journal of Physics: Condensed Matter* 16 (46): 8251–8266. https://doi.org/10.1088/0953-8984/16/46/012.

# Chapter 10
# Conclusions and Perspectives

This thesis presented the development of a method that can describe the dynamics of quantum problems with interacting electrons and phonons. The method, called Effective Correlated Electron-Ion Dynamics (ECEID), aims to describe in real-time the mutual evolution of electrons and phonons in mesoscale systems and capture the energy exchanges between them.

The method was derived in Chap. 5, where, at first, a set of exact equations was retrieved from the model Hamiltonian. Through the application of approximations, a set of equations of motion was determined and implemented in a code. Such a real-time quantum approach is among the more computationally challenging methods for this class of problems, but it is essential for understanding the microscopic processes involved. A key aspect of the method is the linear scaling with the number of vibrational degrees of freedom. This makes it possible to access the typical large size- and time-domains that characterize mesoscale problems. The set of equations conserves total energy on a closed system and can also be used to simulate open systems with the application of open boundaries.

In Chap. 6, ECEID was compared to exact simulations on systems comprising a few degrees of freedom. Under a set of conditions, it was found that the effect of the approximations could be minimized and ECEID converged to the exact case. The application of a bias on a nanowire, with the use of the open boundaries, lets a current flow through the system. Without oscillators, the elastic Landauer two-terminal result was recovered on a perfect nanowire. When oscillators were included in the system, they heated up in the presence of a current, demonstrating the phenomenon of Joule heating. A microscopic Ohm's law could be verified, as the resistance in a nanowire was found to be proportional to the length of the region. The resulting resistivity was validated by comparison with a perturbative result.

Then, in Chap. 7, ECEID was applied to nanowires featuring out-of-equilibrium distributions of electrons and oscillators, to demonstrate its ability to describe thermalization. An entropic definition of temperature, combined with the microscopic ECEID dynamics, produces a thermodynamically meaningful description of the

© Springer International Publishing AG, part of Springer Nature 2018
V. Rizzi, *Real-Time Quantum Dynamics of Electron-Phonon Systems*,
Springer Theses, https://doi.org/10.1007/978-3-319-96280-1_10

energy exchange between the two subsystems, and their equilibration. Thermalization was shown to occur in a range of situations where the commonly used Ehrenfest
dynamics is known to display unphysical results, including a full electronic population inversion. The results were rationalized through a simple model based on rate
equations. The contents of this chapter were published in [1].

In Chap. 8, ECEID was used to perform non-adiabatic quantum simulations of the
inelastic injection and subsequent dynamics of excited electrons in a model water
system. ECEID was applied, at first, to a water molecule between metal leads and
then to a water chain, representing a minimal model of a biological environment.
The inclusion of phonon modes in the water allowed to study elastic and inelastic
effects with the injection of electrons, both as a time dependent pulse and a constant
flow. Part of this chapter has recently been accepted for publication [2].

It was found that the electron-phonon interaction is a key mechanism for the
injection of electrons in water. Inelastic electron-phonon injection is a critical process in granting incoming electrons access to the water excited states. In addition, the
vibrational temperature is a crucial controlling factor in the presence of an incoming electron flux. Depending on the incoming electron energy, the electron-phonon
interaction can activate a current-assisted cooling or an exponential heating, where
the dynamics of the system can further deviate from a steady-state description. By
exploiting the energy-dependent lifetime of the water chain states, the injection of
electrons in specific energy ranges can result in partial electron trapping in the water.
It is hoped that these insights into the dynamical interaction of low energy electrons and water will lead to further more realistic simulations probing DNA damage
mechanisms and to the development of improved models of radiation exposure.

In Chap. 9, a methodological advance was proposed, where the motion of the oscillator centroids was introduced as a time-dependent parameter. The exact dynamics of
the system was derived and, after the application of approximations akin to ECEID,
a set of equations was retrieved, whose limit of zero centroid motion exactly corresponds to ECEID. The new formulation is more efficient to simulate than before and
its performance was tested and compared to the old one. An Ehrenfest-like condition
for the centroid dynamics was proposed and tested.

The new derivation inspires the development of new features in the method. The
directions for improvement of ECEID are numerous and promising. For instance,
the opportunity to go beyond the double (de)excitation approximation was explored
in Appendix D. Following a similar line of reasoning, it would be possible to include
other correlation terms in the EOM, such as correlations between different oscillators.
The full density matrix decoupling looks, for now, the only fundamental approximation that cannot be eliminated at a modest cost.

An essential further development, is the inclusion of anharmonic terms in the
model Hamiltonian, as it would grant a much wider range of applications for the
method, but is far from trivial to implement. Even the simplest anharmonicity, i.e.
a cubic potential, would introduce correlation operators with triplets of creation
and annihilation operators. To devise a consistent set of approximations on such
combinations of operators is no easy task and would require careful thinking.

It is hoped that, in the future, ECEID can be employed on a number of applications and physical problems. One of the most logical extensions is the simulation of 2D or 3D structures. For simplicity, this work is focused on 1D systems such as metallic nanowires or water chains, but the extension to more complex multi-dimensional geometries is natural. In fact, the water simulations already present a 2D structure in the multi-orbital model of water. The idea to use ECEID on large scale simulations of irradiated metallic systems during the electron-phonon equilibration phase is being developed now, in combination with the kinetic model from Sect. 7.4.3.

Recent simulations of radiation damage on metals used Ehrenfest TDDFT [3, 4], but they were intrinsically limited by the Ehrenfest description of the atoms that cannot capture spontaneous phonon emission. Replacing Ehrenfest with ECEID in such simulations would be a significant advancement, but it would not be without serious technical difficulties. A critical dilemma is how to link the atomic motion to the harmonic oscillator formalism of ECEID. The development of ECEID xp is a step forward in the direction of a usable method in combination with TDDFT. The search for a condition to determine the semiclassical dynamics of the oscillator centroids is ongoing and is crucial in providing a link to methods based on classical trajectories.

ECEID can be used to study the interplay between elastic disorder and inelastic scattering on systems where an electronic current flows, as was done in Sect. 6.4.2. A thorough exploration of the parameter space and configurational averaging are required to observe phenomena such as Anderson localization. The appearance of trapped states and their interaction with phonons could also be examined.

ECEID can not only be used to simulate systems with a voltage bias, but also with a temperature bias. The open boundary setup can be modified to include a temperature difference in the probes, so that a thermal injection of electrons and a heat flow can occur. This would allow ECEID to be used for atomistic simulations of thermal transport and thermoelectric effects. This implementation is currently under progress.

The ECEID model Hamiltonian includes phonons as harmonic oscillators. An exciting perspective is to make it include photons too. This would open the opportunity to use ECEID to simulate optical excitations in real time. According to a recent theory [5, 6], another of the five senses involves vibrations: olfaction. Olfaction is seen as a phonon-assisted mechanism based on inelastic electron tunneling. ECEID is already suitable to describe such a model and could represent a useful tool for testing that theory.

We hope that the method derived in this thesis and its applications will give rise to an improved set of methodologies and will inspire their uses in a diverse group of problems.

# References

1. Rizzi, V., T.N. Todorov, J.J. Kohanoff, and A.A. Correa. 2016. Electron-phonon thermalization in a scalable method for real-time quantum dynamics. *Physical Review B* 93 (2): 024306. https://doi.org/10.1103/PhysRevB.93.024306.
2. Rizzi, V., T.N. Todorov, and J.J. Kohanoff. 2017. Inelastic electron injection in a water chain. *Scientific Reports* 7: 45410. https://doi.org/10.1038/srep45410.
3. Correa, A.A., J. Kohanoff, E. Artacho, D. Sánchez-Portal, and A. Caro. 2012. Nonadiabatic forces in ion-solid interactions: The initial stages of radiation damage. *Physical Review Letters* 108 (21): 213201. https://doi.org/10.1103/PhysRevLett.108.213201.
4. Zeb, M.A., J. Kohanoff, D. Sánchez-Portal, A. Arnau, J.I. Juaristi, and E. Artacho. 2012. Electronic stopping power in gold: The role of d electrons and the H/He anomaly. *Physical review letters* 108 (22): 225504. https://doi.org/10.1103/PhysRevLett.108.225504.
5. Brookes, J., F. Hartoutsiou, A. Horsfield, and A. Stoneham. 2007. Could humans recognize odor by phonon assisted tunneling? *Physical Review Letters* 98 (3): 038101. https://doi.org/10.1103/PhysRevLett.98.038101.
6. Brookes, J.C. 2011. Olfaction: The physics of how smell works? *Contemporary Physics* 52 (5): 385–402. https://doi.org/10.1080/00107514.2011.597565.

# Appendix A
# Electronic Operators in ECEID: From Many-Body to Single Body

ECEID's dynamics is determined by a set of EOM (5.34)–(5.41) of scalar quantities and electronic operators. In the general derivation of the method, no assumption is made on the form of the electronic operators that can be in principle as general as one requires. A numerical application of ECEID, though, requires that the operators can be represented by finite size matrices and can be stored in the memory of a computer. The full many-body operators would be overwhelmingly large in any real world application, therefore it is necessary to reduce their size for computational reasons. In this appendix we describe the projection process of the Many-Body electronic operators into Single-Body operators from [1] and apply its results to ECEID.

## A.1 Tracing Over the Electrons

In a system with $N_e$ electrons, we take a generic $N_e$-body operator $\hat{a}^{(N_e)}$ and define its n-body projection by tracing it over $N_e - n$ electrons

$$\hat{a}^{(n)} = \frac{N_e!}{(N_e - n)!} \text{Tr}_{e,n+1,\ldots,N_e}\left(\hat{a}^{(N_e)}\right). \tag{A.1}$$

In [1], the authors take the product of two generic $N_e$-body operators and provide an expression for the one-body form of the product operator.

We consider the special case of the product between an $N_e$-body operator $\hat{a}^{(N_e)}$ and a one-body operator $\hat{b}^{(1)}(i)$ that depends only on electron $i$

$$\hat{c}^{(1)}(1) = N_e \text{Tr}_{e,2,\ldots,N_e}\left(\hat{a}^{(N_e)} \hat{b}^{(1)}(i)\right). \tag{A.2}$$

We point out that the previous and the following expressions do not violate the indistinguishability of electrons. The electron 1 appearing in the left hand side of Eq. (A.2) is not by itself any different from the other electrons. Because of the trace on the right hand side, it effectively represents all other $N_e - 1$ electrons.

© Springer International Publishing AG, part of Springer Nature 2018
V. Rizzi, *Real-Time Quantum Dynamics of Electron-Phonon Systems*,
Springer Theses, https://doi.org/10.1007/978-3-319-96280-1

If $i = 1$, the trace applies only to $\hat{a}^{(N_e)}$ and Eq. (A.2) trivially reduces to

$$\hat{c}^{(1)}(1) = \hat{a}^{(1)}(1)\,\hat{b}^{(1)}(1), \tag{A.3}$$

while, if $i \neq 1$

$$\hat{c}^{(1)}(1) = \frac{1}{N_e - 1}\text{Tr}_{e,2}\left(\hat{a}^{(2)}(1,2)\,\hat{b}^{(1)}(2)\right) \tag{A.4}$$

where we introduce a 2-body operator $\hat{a}^{(2)}(1,2)$.

The following commutator expression

$$N_e \sum_i^{N_e} \text{Tr}_{e,2,\dots,N_e}\left(\left[\hat{a}^{(N_e)},\,\hat{b}^{(1)}(i)\right]\right) = \left[\hat{a}^{(1)}(1),\,\hat{b}^{(1)}(1)\right] \tag{A.5}$$

is valid. For $i = 1$ it is obvious, whereas for $i \neq 1$ the permutation under the trace makes terms such as the one from Eq. (A.4) disappear. On the other hand, the anti-commutator expression

$$N_e \sum_i^{N_e} \text{Tr}_{e,2,\dots,N_e}\left(\left\{\hat{a}^{(N_e)},\,\hat{b}^{(1)}(i)\right\}\right) = \left\{\hat{a}^{(1)}(1),\,\hat{b}^{(1)}(1)\right\} + 2\text{Tr}_{e,2}\left(\hat{a}^{(2)}(1,2)\hat{b}^{(1)}(2)\right). \tag{A.6}$$

presents a two-body term besides the trivial one-body one.

## A.2  Tracing the ECEID Many-Body EOM

The key object that describes the electronic dynamics in ECEID is the electronic DM, which is in general an $N_e$-body operator $\hat{\rho}_e^{(N_e)}$. From Eq. (A.1), the one-body DM is defined as

$$\hat{\rho}_e^{(1)}(1) = N_e \text{Tr}_{e,2,\dots,N_e}\left(\hat{\rho}_e^{(N_e)}\right) \tag{A.7}$$

and the two-body DM is

$$\hat{\rho}_e^{(2)}(1,2) = N_e(N_e - 1)\text{Tr}_{e,3,\dots,N_e}\left(\hat{\rho}_e^{(N_e)}\right). \tag{A.8}$$

In ECEID, we write both the electronic Hamiltonian as a sum of one-body terms $\hat{H}_e = \sum_{i=1}^{N_e} \hat{H}_e^{(1)}(i)$ and the electron–phonon coupling operator $\hat{F}_\nu = \sum_{i=1}^{N_e} \hat{F}_\nu(i)$, neglecting electron-electron interaction.

We first apply $N_e \text{Tr}_{e,2,\dots,N_e}$ to the EOM (5.40) and obtain

$$\dot{\rho}_e^{(1)}(t) = \frac{1}{i\hbar}[\hat{H}_e^{(1)}, \hat{\rho}_e^{(1)}(t)] - \frac{1}{i\hbar}\sum_{\nu=1}^{N_o}[\hat{F}_\nu^{(1)}, \hat{\mu}_\nu^{(1)}(t)] \qquad (A.9)$$

$$\dot{N}_\nu(t) = \frac{1}{\hbar M_\nu \omega_\nu}\mathrm{Tr}_e\left(\hat{F}_\nu^{(1)}\hat{\lambda}_\nu^{(1)}(t)\right) \qquad (A.10)$$

where now all electronic operators depend on one electron only (conventionally electron 1), whose dependency we omit. We used property (A.5) to prove Eqs. (A.9)

$$N_e\sum_i^{N_e}\mathrm{Tr}_{e,2,\ldots,N_e}\left([\hat{H}_e^{(1)}(i), \hat{\rho}_e^{(N_e)}(t)]\right) = [\hat{H}_e^{(1)}(1), \hat{\rho}_e^{(1)}(1,t)], \qquad (A.11)$$

and (A.10)

$$\sum_i^{N_e}\mathrm{Tr}_{e,1,\ldots,N_e}\left(\hat{F}_\nu^{(1)}(i)\hat{\lambda}_\nu(t)\right) = \mathrm{Tr}_e\left(\hat{F}_\nu^{(1)}(1)\hat{\lambda}_\nu^{(1)}(1,t)\right). \qquad (A.12)$$

Equations (A.9) and (A.10) are the natural one-body version of their many-body counterpart. The same happens for Eq. (5.34) and (5.35)

$$\hat{\mu}_\nu^{(1)}(t) = \frac{1}{M_\nu \omega_\nu}(i\,\hat{C}_\nu^{c\,(1)}(t) - \hat{A}_\nu^{s\,(1)}(t)) \qquad (A.13)$$

$$\hat{\lambda}_\nu^{(1)}(t) = i\hat{C}_\nu^{s\,(1)}(t) + \hat{A}_\nu^{c\,(1)}(t). \qquad (A.14)$$

For Eq. (5.36)–(5.39)

$$\dot{\hat{C}}_\nu^{c\,(1)}(t) = -\frac{i}{\hbar}[\hat{H}_e^{(1)}, \hat{C}_\nu^{c\,(1)}(t)] + \omega_\nu \hat{C}_\nu^{s\,(1)}(t) + (N_\nu(t) + \tfrac{1}{2})[\hat{F}_\nu^{(1)}, \hat{\rho}_e^{(1)}(t)] \quad (A.15)$$

$$\dot{\hat{C}}_\nu^{s\,(1)}(t) = -\frac{i}{\hbar}[\hat{H}_e^{(1)}, \hat{C}_\nu^{s\,(1)}(t)] - \omega_\nu \hat{C}_\nu^{c\,(1)}(t) \qquad (A.16)$$

$$\dot{\hat{A}}_\nu^{c\,(1)}(t) = -\frac{i}{\hbar}[\hat{H}_e^{(1)}, \hat{A}_\nu^{c\,(1)}(t)] + \omega_\nu \hat{A}_\nu^{s\,(1)}(t) + \frac{1}{2}\{\hat{F}_\nu, \hat{\rho}_e(t)\}^{(1)} \qquad (A.17)$$

$$\dot{\hat{A}}_\nu^{s\,(1)}(t) = -\frac{i}{\hbar}[\hat{H}_e^{(1)}, \hat{A}_\nu^{s\,(1)}(t)] - \omega_\nu \hat{A}_\nu^{c\,(1)}(t), \qquad (A.18)$$

we employed property (A.5) in the commutator term of Eq. (A.15), while the anti-commutator term in Eq. (A.17) requires further work.

By using property (A.6), it can be written as

$$\frac{1}{2}\{\hat{F}_\nu, \hat{\rho}_e(t)\}^{(1)} = \frac{1}{2}\{\hat{F}_\nu^{(1)}(1), \hat{\rho}_e^{(1)}(1,t)\} + \mathrm{Tr}_{e,2}\left(\hat{F}_\nu^{(1)}(2)\hat{\rho}_e^{(2)}(1,2,t)\right), \quad (A.19)$$

where the first term is intuitive and the second one is more complicated and involves the two-body electronic DM. We do not evaluate the two-body density matrix in ECEID, therefore we need to make an assumption and write it in terms of the quatities that we have. The simplest choice is to write it as a Slater determinant, an antisymmetric linear combination of one-body DM

$$\hat{\rho}_e^{(2)}(12, 1'2') = \hat{\rho}_e^{(1)}(11')\hat{\rho}_e^{(1)}(22') - \hat{\rho}_e^{(1)}(12')\hat{\rho}_e^{(1)}(21'). \tag{A.20}$$

Applying Eq. (A.20) in the second term of Eq. (A.19), we obtain

$$\text{Tr}_{e,2}\left(\hat{F}_\nu^{(1)}(2)\,\hat{\rho}_e^{(2)}(1, 2, t)\right) \simeq \hat{\rho}_e^{(1)}(1, t)\,\text{Tr}_e\left(\hat{F}_\nu^{(1)}(1)\,\hat{\rho}_e^{(1)}(1, t)\right)$$
$$-\hat{\rho}_e^{(1)}(1, t)\,\hat{F}_\nu^{(1)}(1)\,\hat{\rho}_e^{(1)}(1, t). \tag{A.21}$$

With this assumption, Eq. (A.17) becomes

$$\dot{\hat{A}}_\nu^{c\,(1)}(t) = -\frac{i}{\hbar}\,[\hat{H}_e^{(1)}, \hat{A}_\nu^{c\,(1)}(t)] + \omega_\nu \hat{A}_\nu^{s\,(1)}(t) + \frac{1}{2}\,\{\hat{F}_\nu^{(1)}, \hat{\rho}_e^{(1)}(t)\}$$
$$+ \hat{\rho}_e^{(1)}(t)\,\text{Tr}_e\left(\hat{F}_\nu^{(1)}\,\hat{\rho}_e^{(1)}(t)\right) - \hat{\rho}_e^{(1)}(t)\,\hat{F}_\nu^{(1)}\,\hat{\rho}_e^{(1)}(t). \tag{A.22}$$

The one-body EOM involve only one-electron operators, so it is natural and unambiguous to drop the one-body notation as we do in the main text of this thesis.

As explained in Sect. 5.5, in the applications we ignore the term in Eq. (A.22) $\hat{\rho}_e(t)\,\text{Tr}_e(\hat{F}_\nu\hat{\rho}_e(t))$ that is related to the motion of the oscillator centroid. We recover that term in Chap. 9, in the more general derivation of ECEID xp where the motion of the oscillator centroids is reintroduced. In fact, that term is crucial when comparing term by term ECEIDxp with and exact expansion from CEID [1]. A more complex version of the approximation described here is presented in [2] under the name of the *extended Hartree-Fock approximation*.

In this work we always consider spin-degenerate systems. To take it into account, every quantity that results from a total trace over the electrons, gains an extra factor of 2.

# Appendix B
# Open Boundaries in ECEID

We introduce an Open Boundaries (OB) mechanism that allows electron injection and extraction into the system and can be implemented in ECEID to simulate the dynamics of open systems. The general form of the OB method is described in detail in [3].

## B.1    General Formalism

As shown in Fig. 5.2, we take a system (S) and split it into three regions: left (L) and right (R), which can be metal leads, and a central region (C), that is the device region. The device is the item that one wants to investigate and can be, for example, a molecule. It can include phonons. The full Hamiltonian of such a system is

$$\hat{H}_S = \underbrace{\hat{H}_L + \hat{H}_C + \hat{H}_R}_{\hat{H}_{S_0}} + \hat{V} \tag{B.1}$$

where $\hat{H}_L$, $\hat{H}_R$ and $\hat{H}_C$ denote the isolated Hamiltonian of each of the system's subregions, $\hat{H}_{S_0}$ their sum, which corresponds to a system with the leads disconnected from the central region, and $\hat{V}$ the coupling between the central region $C$ and the leads $L$, $R$. We use a one-electron orthonormal atomic basis $|i\rangle$ throughout and call $\hat{I}_N = \sum_{i \in N} |i\rangle\langle i|$ the identity operator over regions $N = C$, $L$ or $R$. We define $\hat{A}_{NN'} = \hat{I}_N \hat{A} \hat{I}_{N'}$ for a generic operator $\hat{A}$ and $\hat{A}_N = \hat{I}_N \hat{A} \hat{I}_N$. Index M here identifies the sum over the left and right leads basis $\hat{I}_M = \sum_{i \in L} |i\rangle\langle i| + \sum_{i \in R} |i\rangle\langle i|$.

A number of sites in the leads are connected to external probes $P_i$ by a constant weak coupling $\gamma_i$. The probes in isolation feature a retarded (advanced) surface Green's function (GF) $\mathfrak{g}_{P_i}^{+(-)}(E)$ and a local density of states $d_{P_i}(E) = -\mathrm{Im}(\hat{\mathfrak{g}}_{P_i}^+(E))/\pi$. The probes in the left and the right lead have electronic distributions $f_{L,R}(E)$.

© Springer International Publishing AG, part of Springer Nature 2018
V. Rizzi, *Real-Time Quantum Dynamics of Electron-Phonon Systems*,
Springer Theses, https://doi.org/10.1007/978-3-319-96280-1

The system's self energy due to the external probes is

$$\hat{\Sigma}^{\pm} = \hat{\Sigma}_L^{\pm} + \hat{\Sigma}_R^{\pm} = \sum_{i \in L} \gamma_i^2 \, \mathfrak{g}_{P_i}^{\pm}(E)) |i\rangle\langle i| + \sum_{i \in R} \gamma_i^2 \, \mathfrak{g}_{P_i}^{\pm}(E)) |i\rangle\langle i| \tag{B.2}$$

and the full system's advanced (retarded) GF can be written as

$$\hat{G}_S^{\pm}(E) = \left( E\hat{I}_S - \hat{H}_S - \hat{\Sigma}^{\pm}(E) \right)^{-1}. \tag{B.3}$$

By solving the Lippman-Schwinger equation for electrons flowing from the probes to the system, the steady state one-body DM is [3]

$$\hat{\rho}_S = \int_{-\infty}^{\infty} \hat{G}_S^{+}(E) \hat{\Sigma}^{<}(E) \hat{G}_S^{-}(E) dE \tag{B.4}$$

where

$$\hat{\Sigma}^{<}(E) = f_L(E) \sum_{i \in L} \gamma^2 d_{P_i}(E) |i\rangle\langle i| + f_R(E) \sum_{i \in R} \gamma^2 d_{P_i}(E) |i\rangle\langle i|. \tag{B.5}$$

To make the picture simpler, we choose to work in the wideband limit where $\gamma_i = \gamma$ does not depend on atomic sites and the probes local density of states is a constant $d_{P_i}(E) = d$. These assumptions make the probes' GF a constant $\mathfrak{g}_{P_i}^{+}(E) = -id\pi$ and Eqs. (B.2), (B.5) become

$$\hat{\Sigma}^{\pm} = \mp i \frac{\Gamma}{2} \left( \hat{I}_L + \hat{I}_R \right) \tag{B.6}$$

$$\hat{\Sigma}^{<}(E) = \frac{\Gamma}{2\pi} \left( f_L(E) \hat{I}_L + f_R(E) \hat{I}_R \right) \tag{B.7}$$

where $\Gamma = 2\pi\gamma^2 d$.

The OB setup is a controlled approximation of the conventional two terminal setup. The metallic leads allow electron injection and extraction into the central region and are instrumental for driving the system away from the initial state. Ideally the details of the leads should not have an influence on the final state that the system reaches at long times. The final state should be determined by the external probes characteristics, such as their population distributions and their chemical potential.

The multiple probe setup allows to hide the finite size of the leads and their discrete level spacing by mimicking the continuous band of an extended system. In the limit of long leads and small $\Gamma$, the leads converge to the semi-infinite case, where the conventional Landauer conduction picture is valid [3]. For this limit to be valid, $\Gamma$ must be larger than the average energy spacing in the leads so that it can effectively hide the finiteness of the leads and their discrete energy spacing. Furthermore, it cannot be too large and has to be smaller than the energy scale at which the electronic structure of a wire changes significantly. A possible upper limit for $\Gamma$ in our simulations is the hopping between sites in the metallic nanowire.

There is freedom in choosing the form of $f_{L,R}$. A typical choice is picking Fermi-Dirac distributions with temperature $T_{L,R}$ and chemical potential $\mu_{L,R}$. For example, a bias $V = \mu_L - \mu_R$ makes an electric current run through the system, possibly driving it to a steady state. Without any inelastic scatterer in the central region, the Landauer two-terminal picture is recovered in the double limit of long leads and small $\Gamma$ but not smaller than the leads level spacing. In a perfect one-dimensional wire with a 1 V bias, the steady state current tends to 77.48 $\mu A$, corresponding to the conductance quantum.

## B.2  Elastic Transmission

A useful quantity that we employ in our simulations is the elastic transmission. To derive a form of it, we consider the GF of the system with the leads disconnected from region C

$$\hat{g}^{\pm}(E) = (E - \hat{H}_{S_0} - \hat{\Sigma}^{\pm})^{-1}. \tag{B.8}$$

By using the Dyson equation, we derive the full system's GF projected on the central region

$$\hat{G}_C^{\pm}(E) = \hat{g}_C^{\pm}(E) + \hat{g}_C^{\pm}(E)\hat{V}_{CM}\hat{G}_{MC}^{\pm}(E) \tag{B.9}$$

$$= \hat{g}_C^{\pm}(E) + \hat{g}_C^{\pm}(E)\hat{\sigma}_C^{\pm}(E)\hat{G}_C^{\pm}(E) \tag{B.10}$$

where we defined the leads self-energy $\hat{\sigma}_C^{\pm}(E) = \hat{V}_{CM}\hat{g}_M^{\pm}(E)\hat{V}_{MC}$. $\hat{G}_C^{\pm}(E)$ can also be written as

$$\hat{G}_C^{\pm}(E) = (E - \hat{H}_C - \hat{\sigma}_C^{\pm})^{-1}. \tag{B.11}$$

The elastic transmission of the system can be written as [4]

$$T(E) = 4\pi^2 \mathrm{Tr}(\hat{t}^{\dagger}(E)\hat{d}_L(E)\hat{t}(E)\hat{d}_R(E)) \tag{B.12}$$

where

$$\hat{t}(E) = \hat{V}_{MC}\hat{G}_C^{+}(E)\hat{V}_{CM} \tag{B.13}$$

and the density of states operator in the leads (isolated from the central region) is

$$\hat{d}_{L,R}(E) = \frac{1}{2\pi i}(\hat{g}_{L,R}^{-}(E) - \hat{g}_{L,R}^{+}(E)). \tag{B.14}$$

Equation (B.12) is a purely elastic static quantity and does not include any inelastic contribution from the phonons. It measures the effect of the system's electronic structure and geometry on the electronic transmission. Its energy dependence can offer insights to the ECEID inelastic calculations, as we see in Chap. 8.

A semiclassical way to include the *elastic* scattering of a phonon $\nu$ is to add an electron-phonon term $\hat{F}_\nu X_\nu$ to $\hat{H}_S$, where $X_\nu$ is a classical coordinate that acts as a parameter. The resulting $T(E, X_\nu)$ represents the probability of an electron with energy $E$ to cross the central region where phonon $\nu$ is statically displaced by $X_\nu$. It is analogous to altering the geometry of the system by using the phonon as a static impurity.

## B.3    Including the OB in ECEID

Consider the Hamiltonian

$$\hat{H} = \hat{H}_S + \hat{H}_P + \hat{H}_{SP} + \hat{H}_{PS} \tag{B.15}$$

describing a system S coupled to external probes P, where $\hat{H}_S$ is a time-independent one-body Hamiltonian for the isolated system, $\hat{H}_P$ represents the probes, $\hat{H}_{SP}$ is the probes-system coupling and $\hat{H}_{SP} = \hat{H}_{PS}^\dagger$. The dynamics of the full system DM is described by the Liouville equation [3, 5]

$$i\hbar\dot{\hat{\rho}}_S(t) = [\hat{H}_S, \hat{\rho}_S(t)] + \hat{H}_{SP}\hat{\rho}_{PS}(t) - \hat{\rho}_{SP}(t)\hat{H}_{PS}. \tag{B.16}$$

where $\hat{H}_{SP}\hat{\rho}_{PS}(t) - \hat{\rho}_{SP}(t)\hat{H}_{PS}$ represent the OB driving terms.

Following the derivation in [3], with the imposition of the wideband limit in the external probes, Eq. (B.16) becomes

$$i\hbar\dot{\hat{\rho}}_S(t) = [\hat{H}_S, \hat{\rho}_S(t)] + \hat{\Sigma}^+\hat{\rho}_S(t) - \hat{\rho}_S(t)\hat{\Sigma}^- + \int_{-\infty}^{\infty} \left( \hat{\Sigma}^<(E)\hat{G}_S^-(E) - \hat{G}_S^+(E)\hat{\Sigma}^<(E) \right) dE \tag{B.17}$$

where $\hat{\Sigma}^+\hat{\rho}_e(t) - \hat{\rho}_e(t)\hat{\Sigma}^-$ describes electron extraction and the energy integral portrays electron injection. $\hat{G}_S^\pm(E) = (E\hat{I}_S - \hat{H}_S \pm i\hat{I}_M\Gamma/2)^{-1}$ is the system's GF.

ECEID's dynamics of a closed system involves the time evolution of $\hat{\rho}_e(t)$ (5.5), $N(t)$ (5.41) and $(\hat{C}_\nu^c, \hat{A}_\nu^c, \hat{C}_\nu^s, \hat{A}_\nu^s)$ (5.36)–(5.39). After connecting the system to the external probes, we assume that the resulting OB driving terms affect only in the EOM of $\hat{\rho}_e(t)$, that turns into

$$i\hbar\dot{\hat{\rho}}_e(t) = [\hat{H}_e, \hat{\rho}_e(t)] - \sum_{\nu=1}^{N_o} [\hat{F}_\nu, \hat{\mu}_\nu(t)] + \hat{\Sigma}^+\hat{\rho}_e(t) - \hat{\rho}_e(t)\hat{\Sigma}^-$$

$$+ \int_{-\infty}^{\infty} \left( \hat{\Sigma}^<(E)\hat{G}_S^-(E) - \hat{G}_S^+(E)\hat{\Sigma}^<(E) \right) dE. \tag{B.18}$$

In next section, we show the procedure that we use to solve the injection integral in a special case.

## B.4  Imposing a Constant Bias in the Leads

The electron injection energy integral

$$\int_{-\infty}^{\infty} \left( \hat{\Sigma}^<(E)\hat{G}_S^-(E) - \hat{G}_S^+(E)\hat{\Sigma}^<(E) \right) dE \tag{B.19}$$

from Eq. (B.18) can be solved analytically in some special cases. For example, if we impose that the probes have zero temperature Fermi-Dirac distributions

$$f_{L,R}(E) = \begin{cases} 1 & \text{if } E \leq \mu_{L,R} \\ 0 & \text{otherwise.} \end{cases} \tag{B.20}$$

We apply a constant bias between the left and right chemical potentials $V = \mu_L - \mu_R$. Inserting $\hat{\Sigma}^{\pm}$ (B.7) in the energy integral (B.19), we have

$$\frac{\Gamma}{2\pi} \int_{-\infty}^{\mu_L} dE \left( \hat{I}_L \hat{G}_S^-(E) - \hat{G}_S^+(E)\hat{I}_L \right) + \frac{\Gamma}{2\pi} \int_{-\infty}^{\mu_R} dE \left( \hat{I}_R \hat{G}_S^-(E) - \hat{G}_S^+(E)\hat{I}_R \right) \tag{B.21}$$

To solve the integrals, we write the full system's retarded GF (B.3) in terms of its eigenstates

$$\hat{G}_S^-(E) = \sum_i \frac{1}{E - \lambda_i} |i_R\rangle\langle i_L| \tag{B.22}$$

where $\lambda_i$ is the eigenvalue corresponding to eigenstate $i$. We point out that $\hat{G}_S^-(E)$ is a non-hermitian operator, so it has a left $|i_L\rangle$ and right $|i_R\rangle$ eigenvector basis. Analogously, the advanced GF is

$$\hat{G}_S^+(E) = \sum_i \frac{1}{E - \lambda_i^*} |i_L\rangle\langle i_R|, \tag{B.23}$$

where $\lambda_i^*$ is the complex conjugate of $\lambda_i$.

We now focus on the integral on the left lead (the first term in Eq. (B.21)) and solve it on the system's atomic basis.[1] It can be written as

$$\frac{\Gamma}{2\pi} \int_{-\infty}^{\mu_L} dE \sum_{i,L} \left( \frac{1}{E - \lambda_i} |L\rangle\langle L|i_R\rangle\langle i_L| - \frac{1}{E - \lambda_i^*} |i_L\rangle\langle i_R|L\rangle\langle L| \right). \tag{B.24}$$

---

[1] The right lead term can be recovered with a similar derivation.

After projecting it on the full system's atomic basis, it becomes

$$\sum_{S,S'} \frac{\Gamma}{2\pi} |S\rangle\langle S'| \sum_i \int_{-\infty}^{\mu_L} dE \left( \frac{c'_{S,S',i}}{E - \lambda_i} - \frac{c_{S,S',i}}{E - \lambda_i^*} \right) \tag{B.25}$$

where we defined the coefficients

$$c'_{S,S',i} = \sum_L \langle S|L\rangle\langle L|i_R\rangle\langle i_L|S'\rangle \tag{B.26}$$

$$c_{S,S',i} = \sum_L \langle S|i_L\rangle\langle i_R|L\rangle\langle L|S'\rangle. \tag{B.27}$$

The solution of the first term in Eq. (B.25) is

$$\sum_{S,S'} \frac{\Gamma}{2\pi} |S\rangle\langle S'| \sum_i c'_{S,S',i} \left( \frac{1}{2} \log\left( \frac{(\mu_L - \Re(\lambda_i))^2 + \Im(\lambda_i)^2}{(B - \Re(\lambda_i))^2 + \Im(\lambda_i)^2} \right) + i\left( \arctan\frac{\mu_L - \Re(\lambda_i)}{\Im(\lambda_i)} + \frac{\pi}{2} \right) \right) \tag{B.28}$$

where $\Re(\lambda_i)$ is the real part of $\lambda_i$ and $\Im(\lambda_i)$ is its imaginary part. For computational purposes, when necessary we replaced the lower limit of the integral $-\infty$ with an arbitrarily negative energy cutoff $B$. The convergence in $B$ was verified.

It is now straightforward to solve the full injection integral (B.21). Its solution on the atomic basis is

$$\sum_{S,S'} \frac{\Gamma}{2\pi} |S\rangle\langle S'| \quad \sum_i \left( \left( c'_{S,S',i} - c_{S,S',i} \right) \frac{1}{2} \log\left( \frac{(\mu_L - \Re(\lambda_i))^2 + \Im(\lambda_i)^2}{(B - \Re(\lambda_i))^2 + \Im(\lambda_i)^2} \right) \right.$$

$$+ i\left( c'_{S,S',i} + c_{S,S',i} \right)\left( \arctan\frac{\mu_L - \Re(\lambda_i)}{\Im(\lambda_i)} + \frac{\pi}{2} \right)$$

$$+ \left( d'_{S,S',i} - d_{S,S',i} \right)\frac{1}{2} \log\left( \frac{(\mu_R - \Re(\lambda_i))^2 + \Im(\lambda_i)^2}{(B - \Re(\lambda_i))^2 + \Im(\lambda_i)^2} \right)$$

$$\left. + i\left( d'_{S,S',i} + d_{S,S',i} \right)\left( \arctan\frac{\mu_R - \Re(\lambda_i)}{\Im(\lambda_i)} + \frac{\pi}{2} \right) \right) \tag{B.29}$$

where, in an analogous way for the right lead, we defined

$$d'_{S,S',i} = \sum_R \langle S|R\rangle\langle R|i_R\rangle\langle i_L|S'\rangle \tag{B.30}$$

$$d_{S,S',i} = \sum_R \langle S|i_L\rangle\langle i_R|R\rangle\langle R|S'\rangle. \tag{B.31}$$

Equation (B.29) can be inserted in the EOM for $\hat{\rho}_e(t)$ (B.18) to allow electron injection. The inclusion of injection does not have a negative impact on the simulation performance because it is a time-independent quantity that can be calculated before the start of the ECEID dynamics and stored.

## B.5 Other Injection Setups

Other than the zero temperature bias setup presented in the previous section, it is possible to vary the probes' electronic distributions $f_{L,R}(E)$ to allow other injection setups.

In Chap. 8, we introduce the electron-gun in which we inject streams of electrons in the system within a narrow energy window. We start from equilibrium OB where the probes are kept at zero electronic temperature and their bias is set to zero. On top of that, $f_L(E)$ contains a top-hat spike between $E_D = \epsilon - \delta\epsilon$ and $E_U = \epsilon + \delta\epsilon$, to allow electrons in that energy interval to be injected in the system. In this case, the injection integral (B.29) gains these additional terms

$$\sum_{S,S'} \frac{\Gamma}{2\pi} |S\rangle\langle S'| \quad \sum_i \left( \left( c'_{S,S',i} - c_{S,S',i} \right) \frac{1}{2} \log\left( \frac{(E_U - \Re(\lambda_i))^2 + \Im(\lambda_i)^2}{(E_D - \Re(\lambda_i))^2 + \Im(\lambda_i)^2} \right) \right.$$

$$\left. + i\left( c'_{S,S',i} + c_{S,S',i} \right) \left( \arctan \frac{E_U - \Re(\lambda_i)}{\Im(\lambda_i)} - \arctan \frac{E_D - \Re(\lambda_i)}{\Im(\lambda_i)} \right) \right). \quad \text{(B.32)}$$

Another electron injection setup that we can implemented is the inclusion of a temperature difference between the left and right probes. This allows simulations of thermoelectric phenomena. For finite temperatures in the probes, the injection integral cannot be determined analytically and has to be solved numerically.

A recent generalization of the OB method in the static limit is presented in [6]. The authors also introduce an innovative mechanism that allows to reduce the length of the leads by using a local coupling to the probes $\Gamma_i$.

# Appendix C
# An Alternative Water Chain

In this appendix, we use a different water chain model that represents an intermediate step between the water molecule in Sect. 8.1 and the water chain in 8.2. This simplified model employs hoppings akin to the ones of the water molecule. The simulations display a curious localization effect that we did not observe in the more complex water chain model and is determined by the peculiar band structure.

The water chain equilibrium geometry shown in Fig. 8.14 is unchanged and the bond lengths $R_{OH_1}$, $R_{OH_2}$, $R_{OO}$ and $\beta$ are the same as in Sect. 8.2. The Hamiltonian for the $j$th water molecule has again a form

$$\hat{H}_j = \begin{bmatrix} E_{H_1} - E_F & \mp W_1 \cos\theta & W_1 \sin\theta & 0 \\ \mp W_1 \cos\theta & E_{O_{p_x}} - E_F & 0 & \pm W_2 \cos\theta \\ W_1 \sin\theta & 0 & E_{O_{p_z}} - E_F & W_2 \sin\theta \\ 0 & \pm W_2 \cos\theta & W_2 \sin\theta & E_{H_2} - E_F \end{bmatrix}$$

where now the O-H hopping have this form $W(R) = 1.84 \frac{\hbar^2}{4 m_e R^2}$, just as in the water molecule section. The inter-orbital hoppings are $W_1 = W(R_{OH_1}) = 3.72$ eV and $W_2 = W(R_{OH_2}) = 3.51$ eV and the inter-molecular hopping is $W(R_{OO} - R_{OH_2}) = 1.26$ eV, much higher than in Sect. 8.2, mainly because of the absence of a cutoff distance.

The phonon mode geometry is the same as in the previous section, with the vibrating $O - H_1$ bonds, as shown in Fig. 8.14, with the same frequency, mass and form of $\hat{F}$. The electron-phonon coupling is different (the hoppings changed) and is
$$F = C \frac{\partial W(R)}{\partial R}\bigg|_{R=R_{OH_1}} = 2.43 \text{ eV/Å, with } C = 1/\sqrt{10}.$$

## C.1  Simulation Details

The eigenvalues of a water chain of 10 molecules form 4 bands, where the bandgap between the lower and the upper ones (the FCB and the SCB) is 4.27 eV.

© Springer International Publishing AG, part of Springer Nature 2018
V. Rizzi, *Real-Time Quantum Dynamics of Electron-Phonon Systems*,
Springer Theses, https://doi.org/10.1007/978-3-319-96280-1

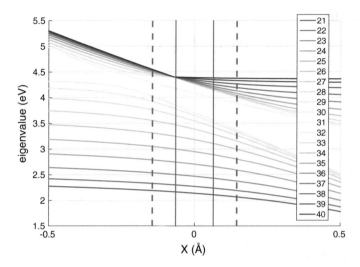

**Fig. C.1** Water chain eigenvalues and their elastic variation with the vibrational displacement $X$ in the frozen phonon regime. The zero point motion of the phonon is indicated by the solid vertical lines, while the root mean square displacement for $N = 2$ is shown with dashed vertical lines

By scanning in $X$, we explore the elastic variation of the eigenvalues and show it in Fig. C.1. In this case, the gap between the FCB and the SCB is much narrower than before. The FCB has a shape resembling the one in Fig. 8.16, with a larger width due to the larger inter-molecular hopping. The SCB presents a degenerate point where all eigenvalues but one converge in energy. Eigenstate 31 is the only one behaving differently. To understand better what makes that state differ from the other ones, we project it on the atomic orbital basis and compare it with other eigenstates in Fig. C.2.

The square modulus projection of eigenstate 21 reaches its maximum for the molecules in the middle of the chain and smoothly decreases moving towards the borders of the chain. Eigenstate 22 has a node in the middle of the chain and two maxima to its left and its right. These projections mainly involve the $H_2$ and the $O\,2p_z$ orbitals. They are analogous to the wavefunctions of a particle in a box, as they present an increasing number of nodes as their energy increases.

The next eigenstates show a similar behaviour, except for the states close to the upper edge of the band. For example, eigenstate 30 follows a different pattern: it lies mainly in the first half of the chain. It does not present any sizeable projection on water molecule number 10 and lies mainly on $H_1$. On the other hand, eigenstate 31 has a complementary form, with its projection being almost exclusively localized on molecule 10. With eigenstate 32, the trend of projections resembling wavefunctions of a particle in a box reappears and carries on until eigenstate 40, with the dominant orbitals for eigenstate 32 being $H_1$ and $O\,2p_x$, while for eigenstate 40 being $H_2$.

The elastic transmission of this system, shown in Fig. C.3, presents peaks corresponding to the eigenvalues of the FCB and SCB, reproducing the eigenvalue scenario from Fig. C.1. The introduction of metal leads is analogous to the procedure

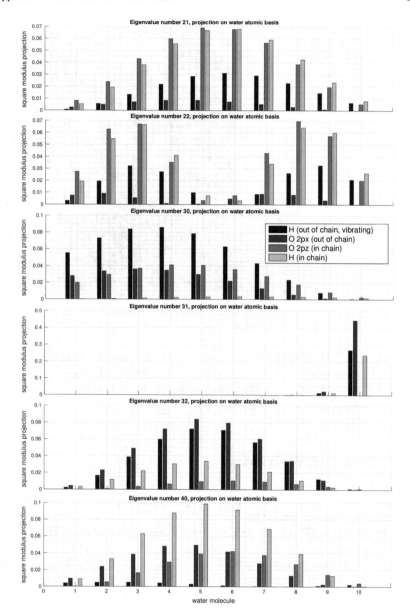

**Fig. C.2** Square modulus of the projection of a number of eigenstates on the atomic orbital basis. Dark blue peaks correspond to the hydrogen orbital sticking out of the chain $H_1$ (the vibrating one), light blue to the oxygen $2p_x$, aquamarine to the oxygen $2p_y$ and yellow to the hydrogen pointing in the chain $H_2$. The peaks form 10 groups that correspond to the 10 water molecules in the chain, in order from left to right. From top to bottom, water chain eigenstates 21, 22, 30, 31, 32, 40 are shown

**Fig. C.3** Elastic
transmission of the FCB and
SCB of a 10 molecule water
chain, with a phonon
displacement range
$-0.5\text{Å} < X < 0.5\text{Å}$

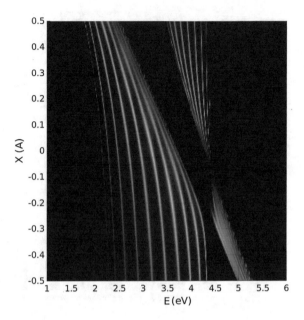

in Sect. 8.2, with 80 metallic sites attached to the left and to the right of the chain,
a metal-chain hopping of $w = 1.0$ eV. The metal has a nearest neighbour hopping
$t = -4$ eV and metallic onsite energies of 8 eV. The 40 leftmost and rightmost lead
atoms are connected to external probes by $\Gamma = 1.25$ eV.

## C.2    Electron-Pulse Injection

As in Sect. 8.2.3, the left lead contains a wavepacket (whose form is in Eq. 8.10) that
propagates to the right until it collides with the water chain. Here the central site
of the pulse is $n_0 = 50$ and its width is $\sigma = 10$ sites. The phonon is initialized at
$N(0) = 0$ or 2.

We inject electrons in both the FCB and SCB by scanning over the wavepacket
momentum $k$, which corresponds to a scan in average energy of the injected
wavepacket $\langle E \rangle$. Here we call EEP the sum over the occupation of the all the water
states in both the FCB and the SCB. It is shown at different times in Fig. C.4a. 5
fs after the simulation start, the EEP displays a peak in the energy bands and is
remarkably nonzero for pulses both below and above the water band edges, as we
observed in the previous water chain simulations in Sect. 8.2. A feature that did not
appear previously is the energy shift of the peak for long times. As time progresses,
the EEP peak decreases in magnitude and gradually drifts towards the upper edge of
the SCB. This band edge effect will be analysed later in more detail. The EEP does
not present marked differences between $N(0) = 0$ and $N(0) = 2$.

**Fig. C.4** Electron-pulse
simulations where Gaussian
wavepackets collide with the
water chain. In **a**, EEP in the
water chain for wavepackets
with average energy $\langle E \rangle$ at 5
fs (blue), 50 fs (red), 100 fs
(green), 200 fs (yellow), 500
fs (purple) after the start of
the simulation. The phonon
starts at $N(0) = 0$ for dashed
lines, at $N(0) = 2$ for solid
lines. The vertical black lines
mark the energy range of the
water chain empty
eigenstates (the lower FCB
and the upper SCB edges). In
**b**, the phonon variation $\Delta N$
in electron-pulse simulations
after 500 fs for a starting
$N(0) = 0$ (dashed line) and
$N(0) = 2$ (solid line)

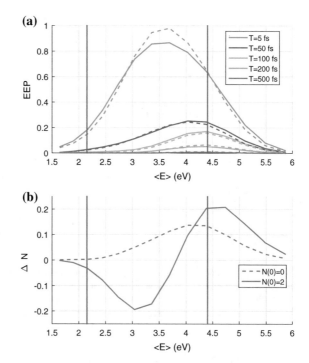

The nonzero EEP in the forbidden regions is determined by the energy spread of
the gaussian pulse and by the inelastic hopping of the electrons, that, thanks to the
presence of the phonon, can gain access to the water band. By checking the phonon
variation $\Delta N(t)$ at $t = 500$ fs in Fig. C.4b, we observe a clearly different behaviour
between different initial $N(0)$, in analogy with the observation from Fig. 8.17.

For pulses below the water band edge $\langle E \rangle < 2.15$ eV, an electron can only absorb
phonons to hop on the band, therefore this inelastic process can be activated only
for $N(0) \neq 0$. In fact, we observe that $\Delta N < 0$ for $N(0) = 2$, while $\Delta N = 0$ for
$N(0) = 0$. Electrons from pulses above the band edge $\langle E \rangle > 4.4$ eV have to emit
phonons while entering the band. In the simulations, both $N(0) = 0$ and $N(0) = 2$
show $\Delta N > 0$ for $\langle E \rangle > 5.5$ eV, with a larger $\Delta N$ for $N(0) = 2$, as the total emission
rate (spontaneous plus stimulated) is larger. The EEP in this case is intense and must
be caused by the inelastic hopping. In contrast with Sect. 8.2, here we have injected
electrons all over the energy range of the unoccupied bands. The inelastic effect when
injecting above the SCB is clearer because of the absence of energy levels above it.

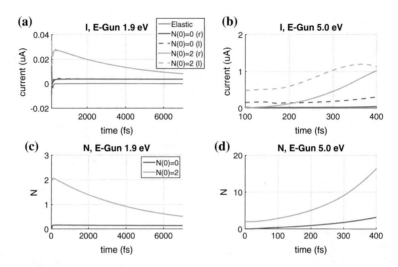

**Fig. C.5** Electron-gun simulations where we inject electrons within a narrow energy window slightly above the SCB and below the FCB. In **a** and **b**, current measured on the left (dashed) and on the right (solid) of the water chain with an injection at $\epsilon = 1.9 \pm 0.1$ eV and $\epsilon = 5.0 \pm 0.1$ eV. In **c** and **d**, dynamics of $N$ during the electron-gun simulations (**a**) and (**b**) respectively

## C.3    Electron-Gun Injection

We perform Electron-gun simulations, just as we did in Sect. 8.2.4. We shoot electrons at $\epsilon = 1.9$ eV, just below the FCB, and at $\epsilon = 5.0$ eV, above the SCB, with a $\Delta\epsilon = 0.1$ eV.

We observe very similar results when comparing to the previous case. The current, that we show in Fig. C.5a and b, is zero in the elastic case, for $\epsilon = 5.0$ eV it increases and shows disagreement between the left and the right of the chain and for $\epsilon = 1.9$ eV it settles down at a small value. In Fig. C.5c and d, we track the dynamics of $N(t)$ that cools down for $\epsilon = 1.9$ eV and heats up for $\epsilon = 5.0$ eV. The same physical phenomena of current assisted cooling and heating occur and the observations from Sect. 8.2.4 are reaffirmed.

## C.4    Eigenstate Lifetime and Bandedge Trapping

We go back to the electron-pulse simulations in Fig. C.4a, where we observed that the absorption peaks gradually moved towards the chain upper bandedge for increasing times. To understand this effect better, we introduce $\Delta$EEP as the percentage variation of the EEP at a given time over its value at $t = 5$ fs and show it in Fig. C.6a. This quantity measures how much of the initial excess electron population remains in the chain as a function of the pulse average energy $\langle E \rangle$ and time.

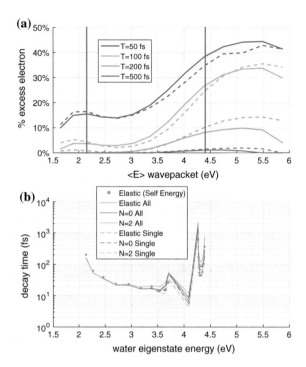

**Fig. C.6** In **a**, electron-pulse ΔEEP, the percentage variation of EEP in the water chain with respect to the value at $t = 5$ fs for $N(0) = 0$ (dashed line) and $N(0) = 2$ (solid line). $t = 50$ fs are in red, $t = 100$ fs in green, $t = 200$ fs in yellow, $t = 500$ fs in purple. In **b**, lifetime comparison of the water chain FCB and SCB eigenstates versus their energy. The solid dots are obtained from static self energy calculations. The solid lines are exponential fits of the decaying population of the single populated level, while the dashed lines are exponential fits of the population of all levels in the FCB and the SCB. Blue curves correspond to elastic real time calculations, red ones to $N(0) = 0$ and green ones to $N(0) = 2$, both with $\dot{N} = 0$

The highest percentage of captured electron occurs for pulses centred about 1 eV above the water SCB. Such pulses also present the longest lived effect on the system. The water highest energy states act as an electron trap. This trapping effect is analogous to the bandedge trapping observed in Sect. 8.2.5. Here it is more evident, as it appears directly in ΔEEP, while previously it appeared only after extracting the inelastic contribution from the EEP.

Just as we did previously, we compute the lifetimes of each water eigenstate $j$ in the FCB and the SCB. We employ self energy calculations $\tau_{\Sigma,j}$ and fits of initially populated eigenstates, on the single initially populated level $\tau_{s,j}$ and on all levels $\tau_{a,j}$. These lifetimes are shown as a function of eigenstate energy in Fig. C.6b. The eigenstates belonging to the SCB are all very close in energy and display much longer lifetimes than the states in the FCB. Therefore, it is intuitive that pulses centred at and above the SCB edge give rise to electrons that are very long lived, as the states that would be mostly populated by them belong to the SCB.

**Fig. C.7** Lifetimes of the
FCB and SCB eigenstates as
in Fig. C.6b, with the
eigenstate number on the x
axis

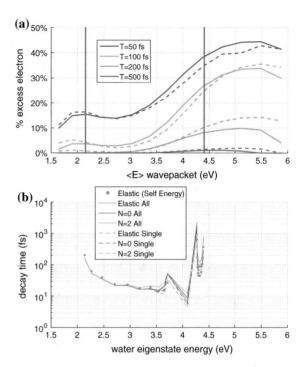

To compare the lifetimes in a clearer way, we plot the data from Fig. C.6b in
Fig. C.7 as a function of the eigenstate number. The characteristic horseshoe shape
that we observed in Sect. 8.2.5 becomes apparent for both the FCB and the SCB, and
the lifetime remarks from there are valid here.

In the FCB (states with $21 \leq j \leq 30$), $\tau_{a,j}$ and $\tau_{s,j}$ closely resemble $\tau_{\Sigma,j}$, indicating that their leading electron escape mechanism is the elastic one to the leads.
Level 31 uniquely presents a much shorter lifetime and is the only eigenstate that
does not fit in the horseshoe shape. As noticed earlier in Fig. C.2, it is the only level
whose projection on the atomic base lies almost exclusively on water molecule 10,
the closest molecule to the right lead. An electron occupying that level has a fast
elastic escape route from the chain to the lead. Levels $32 \leq j \leq 40$ in the SCB show
much longer lifetimes than the ones on the FCB. This difference can be qualitatively
understood by considering the eigenstate wavefunction projection on the chain ends.
In Fig. C.2, for example, the projection on the left end of the chain of eigenstate 30,
belonging to the FCB, is much larger than the one of the SCB states shown there.

As $\tau_{a,j}$ differs more from $\tau_{s,j}$ in the SCB states than for the FCB states, diffusion
to neighbouring levels in the water chain is more relevant in the SCB. The inelastic
cases $N = 0, 2$ decay faster because, thanks to inelastic processes, electrons can
change their energy, eventually gaining access to rapidly decaying states.

# Appendix D
# Beyond the Double (De)excitation Approximation

The new formalism from Chap. 9 is a source of inspiration for ways to go beyond some ECEID approximations. Presented below is a version of ECEID xp that can do without the double (de)excitation approximation.

The derivation starts with the definition[2] of the scalar quantities

$$C_\nu(t) = \text{Tr}\left(\hat{a}_\nu^\dagger \hat{a}_\nu^\dagger \hat{\rho}(t)\right) \tag{D.1}$$

$$D_\nu(t) = \text{Tr}\left(\hat{a}_\nu \hat{a}_\nu \hat{\rho}(t)\right), \tag{D.2}$$

that were previously ignored by approximation.

The exact expressions (9.23) and (9.24) are the only place where the approximations are applied in ECEID xp. By employing the newly defined quantities (D.1 and D.2) and applying the DM decoupling $\hat{\rho}(t) \simeq \hat{\rho}_e(t)\hat{\rho}_o(t)$ and the single oscillator $\text{Tr}_o\left(\hat{a}_\nu^{(\dagger)}\hat{a}_{\nu'}^{(\dagger)}\hat{\rho}(t)\right) = \text{Tr}_o\left(\hat{a}_\nu^{(\dagger)}\hat{a}_{\nu'}^{(\dagger)}\hat{\rho}(t)\right)\delta_{\nu\nu'}$ approximations, the expressions from (9.25) and (9.26) become

$$\text{Tr}_o\left(\hat{\xi}_\nu^2 \hat{\rho}(t)\right) \approx \frac{\hbar}{2M\omega}\left(2N_\nu(t) + 1 + C_\nu(t) + D_\nu(t)\right)\hat{\rho}_e(t) \tag{D.3}$$

$$\text{Tr}_o\left(\hat{\pi}_\nu\hat{\xi}_\nu \hat{\rho}(t)\right) \approx \frac{i\hbar}{2}\left(C_\nu(t) - D_\nu(t) - 1\right)\hat{\rho}_e(t), \tag{D.4}$$

where the new terms are indicated in green.

These quantities appeared in the EOM for the auxiliary operators (9.28) and (9.29) that now can be written as

---

[2] Or, equivalently, $C_\nu(t) = \text{Tr}_o\left(\hat{a}_\nu^\dagger \hat{a}_\nu^\dagger \hat{\rho}_o(t)\right)$ and $D_\nu(t) = \text{Tr}_o\left(\hat{a}_\nu \hat{a}_\nu \hat{\rho}_o(t)\right)$.

© Springer International Publishing AG, part of Springer Nature 2018
V. Rizzi, *Real-Time Quantum Dynamics of Electron-Phonon Systems*,
Springer Theses, https://doi.org/10.1007/978-3-319-96280-1

$$i\hbar\dot{\hat{\mu}}_\nu(t) = [\hat{H}_e, \hat{\mu}_\nu(t)] + \frac{i\hbar}{M}\hat{\lambda}_\nu(t) - \frac{\hbar}{M\omega}\left(N(t) + \frac{1}{2}\right)[\hat{F}_\nu, \hat{\rho}_e(t)]$$

$$- \sum_{\nu'=1}^{N_o}\left(x_{\nu'}(t)[\hat{F}_{\nu'}, \hat{\mu}_\nu(t)]\right) - \frac{\hbar}{2M\omega}\left(C_\nu(t) + D_\nu(t)\right)[\hat{F}_\nu, \hat{\rho}_e(t)] \quad \text{(D.5)}$$

$$i\hbar\dot{\hat{\lambda}}_\nu(t) = [\hat{H}_e, \hat{\lambda}_\nu(t)] - i\hbar K\hat{\mu}_\nu(t) + \frac{i\hbar}{2}\{\hat{F}_\nu, \hat{\rho}_e(t)\} - i\hbar\text{Tr}_e\left(\hat{F}_\nu\hat{\rho}_e(t)\right)\hat{\rho}_e(t)$$

$$- \sum_{\nu'=1}^{N_o}\left(x_{\nu'}(t)[\hat{F}_{\nu'}, \hat{\lambda}_\nu(t)]\right) + \frac{i\hbar}{2}\left(D_\nu(t) - C_\nu(t)\right)[\hat{F}_\nu, \hat{\rho}_e(t)]. \quad \text{(D.6)}$$

The above expressions, together with the unchanged Eqs. (9.10) and (9.13), represent the set of ECEID xp EOM without the use of the double (de)excitation approximation. The novelty lies in the presence of $C_\nu(t)$ and $D_\nu(t)$, that need an EOM themselves to close the set of equations.

Just as in the derivation of $\dot{N}_\nu(t)$ in (9.11), the following time derivatives are valid

$$\dot{C}_\nu(t) = \text{Tr}\left(\hat{a}_\nu^\dagger\hat{a}_\nu^\dagger\dot{\hat{\rho}}(t)\right) = \frac{1}{i\hbar}\text{Tr}\left([\hat{a}_\nu^\dagger\hat{a}_\nu^\dagger, \hat{h}(t)]\hat{\rho}(t)\right) \quad \text{(D.7)}$$

$$\dot{D}_\nu(t) = \text{Tr}\left(\hat{a}_\nu\hat{a}_\nu\dot{\hat{\rho}}(t)\right) = \frac{1}{i\hbar}\text{Tr}\left([\hat{a}_\nu\hat{a}_\nu, \hat{h}(t)]\hat{\rho}(t)\right) \quad \text{(D.8)}$$

where the model Hamiltonian $\hat{h}(t)$ is (9.8). By using the canonical commutation relations and the DM decoupling approximation, the EOM for the new scalar quantities can be written as

$$\dot{C}_\nu(t) = 2\omega i C_\nu(t) - \frac{1}{i\hbar}\left(M\ddot{x}_\nu(t) + Kx_\nu(t)\right)\left(\bar{\xi}_\nu(t) - \frac{i}{M\omega}\bar{\pi}_\nu(t)\right)$$

$$+ \frac{1}{i\hbar}\left(\text{Tr}_e\left(\hat{F}_\nu\hat{\mu}_\nu(t)\right) - \frac{i}{M\omega}\text{Tr}_e\left(\hat{F}_\nu\hat{\lambda}_\nu(t)\right)\right) \quad \text{(D.9)}$$

$$\dot{D}_\nu(t) = -2\omega i D_\nu(t) + \frac{1}{i\hbar}\left(M\ddot{x}_\nu(t) + Kx_\nu(t)\right)\left(\bar{\xi}_\nu(t) + \frac{i}{M\omega}\bar{\pi}_\nu(t)\right)$$

$$- \frac{1}{i\hbar}\left(\text{Tr}_e\left(\hat{F}_\nu\hat{\mu}_\nu(t)\right) + \frac{i}{M\omega}\text{Tr}_e\left(\hat{F}_\nu\hat{\lambda}_\nu(t)\right)\right), \quad \text{(D.10)}$$

where $\bar{\xi}_\nu(t)$ and $\bar{\pi}_\nu(t)$ are defined in (9.16 and 9.17).

The new set of equations still requires a condition for $x_\nu(t)$. That condition can be the Ehrenfest-like (9.36) from ECEID xp, $x_\nu(t) = 0$ that generates ECEID or another one yet to be found.

# References

1. Horsfield, A.P., D.R. Bowler, A.J. Fisher, T.N. Todorov, and C.G. Sánchez. 2004. Beyond Ehrenfest: correlated non-adiabatic molecular dynamics. *Journal of Physics: Condensed Matter* 16 (46): 8251–8266. https://doi.org/10.1088/0953-8984/16/46/012.
2. Horsfield, A.P., D.R. Bowler, A.J. Fisher, T.N. Todorov, and C.G. Sanchez. 2005. Correlated electron-ion dynamics: the excitation of atomic motion by energetic electrons. *Journal of Physics: Condensed Matter* 17 (30): 4793–4812. https://doi.org/10.1088/0953-8984/17/30/006.
3. McEniry, E.J., D.R. Bowler, D. Dundas, A.P. Horsfield, C.G. Sánchez, and T.N. Todorov. 2007. Dynamical simulation of inelastic quantum transport. *Journal of Physics: Condensed Matter* 19 (19): 196201. https://doi.org/10.1088/0953-8984/19/19/196201.
4. Todorov, T.N. 2002. Tight-binding simulation of current-carrying nanostructures. *Journal of Physics: Condensed Matter* 14 (11): 3049–3084. https://doi.org/10.1088/0953-8984/14/11/314.
5. Horsfield, A.P., D.R. Bowler, and A.J. Fisher. 2004. Open-boundary Ehrenfest molecular dynamics: Towards a model of current induced heating in nanowires. *Journal of Physics: Condensed Matter* 16 (7): L65–L72. https://doi.org/10.1088/0953-8984/16/7/L03.
6. Horsfield, A.P., M. Boleininger, R. D'Agosta, V. Iyer, A. Thong, T.N. Todorov, and C. White. 2016. Efficient simulations with electronic open boundaries. *Physical Review B* 94 (7): 075118. https://doi.org/10.1103/PhysRevB.94.075118.

Printed in the United States
By Bookmasters